Textile dyes

Textile dyes

N. N. Mahapatra

WOODHEAD PUBLISHING INDIA PVT LTD

New Delhi

Published by Woodhead Publishing India Pvt. Ltd.
Woodhead Publishing India Pvt. Ltd.,
303, Vardaan House, 7/28, Ansari Road,
Daryaganj, New Delhi - 110002, India
www.woodheadpublishingindia.com

First published 2016, Woodhead Publishing India Pvt. Ltd.
© Woodhead Publishing India Pvt. Ltd., 2016

Woodhead Publishing India Pvt. Ltd. ISBN: 978-93-85059-04-9
Woodhead Publishing India Pvt. Ltd. e-ISBN: 978-93-85059-60-5

Typeset by Mind Box Solutions, New Delhi
Printed and bound by Replika Press Pvt. Ltd.

Contents

Preface

During our college days we used to read a book on "Chemistry of Dyes," written by Prof Dr Venkatraman. That was into many volumes and very big in size. It was a Bible for students studying Dyestuff Chemistry. But the students found it very difficult to understand. Then Dr Shenai wrote little simplified books on dyes and application. We read those books in our college time.

But all these books became very academic. When we entered the textile industry the scenario was different. The practices adopted in dyes units and textile companies were very different.

Till then nobody from the industry wrote any book on dyes and textiles. So I have put my 30-year shop floor experience into the book which I have titled "Textile Dyes."

Each chapter is simplified into the major class of dyes. I have dealt with the history, manufacturing, properties, identification, stripping, testing and application of dyes.

It is written in a very simple language and in a lucid manner.

The book will be helpful to textile students, research students, supervisors working in the dyes and textile industries.

For the last 2 years, the prices of H Acid were so high that the prices of Reactive Dyes went skyrocketing. Every person in the industry was worried for H Acid. I found that time everybody came to know about H Acid. How crucial raw material or dye intermediate it was for the dyes industry. But they didn't know the role of H Acid in manufacturing of reactive dyes. This book deals about the manufacturing process.

This will be helpful both to the dyes industry as well as textile industry.

Acknowledgements

I am thankful to our MD Mr Subhash Bhargava who gave me an opportunity to write such a book. His inspiration was a boost to me to put all my thoughts and studies into a book. My special thanks to my colleague Bharat Trivedi for his help in compiling the chapters in the computer.

Last but not the least my sincere thanks to my family members for their timely help to complete this book in time.

I want to dedicate this book to my Guide and philosopher Late Prof Dr V.A. Shenai by whom I was inspired to write few textile books.

Conclusion

This is my 5th book I have written on textiles. There are some interesting queries answered. People will find it very handy. It will become a reference book for Industry owners, Production Staff and Textile students and Research scholars.

1

Introduction to textile dyes

1.1   Introduction

Colorants are used in many industries – in clothes, paints, plastics, photographs, prints, and ceramics. Colorants are also now being used in novel applications and are termed functional (high technology), as they are not just included in the product for aesthetic reasons but for specific purposes, for example in surgery.

Colorants can be either *dyes or pigments*. Dyes are soluble, coloured organic compounds that are usually applied to textiles from a solution in water. They are designed to bond strongly to the polymer molecules that make up the textile fiber.

Pigments are insoluble compounds used in paints, printing inks, ceramics and plastics. They are applied by using dispersion in a suitable medium. Most pigments used are also organic compounds.

Dyes are soluble at some stage of the application process, whereas pigments, in general, retain essentially their particulate or crystalline form during application. A dye is used to impart colour to materials of which it becomes an integral part. An aromatic ring structure coupled with a side chain is usually required for resonance and thus to impart colour (Resonance structures that cause displacement or appearance of absorption bands in the visible spectrum of light are responsible for colour). Correlation of chemical structure with colour has been accomplished in the synthesis of dye using a chromogen-chromophore with auxochrome. Chromogen is the aromatic structure containing benzene, naphthalene, or anthracene rings. A chromophore group is a colour giver and is represented by the following radicals, which form a basis for the chemical classification of dyes when coupled with the chromogen: azo (–N=N–); carbonyl (=C=O); carbon(=C=C=); carbon-nitrogen (>C=NH or –CH=N–); nitroso (–NO or N–OH); nitro (–NO or =NO–OH); and sulfur (>C=S, and other carbon-sulfur groups). The chromogen-chromophore structure is often not sufficient to impart solubility and cause adherence of dye to fiber. The auxochrome or bonding affinity groups are amine, hydroxyl, carboxyl, and sulfonic radicals, or their derivatives. These auxochromes are important in the use classification of dyes. A listing of dyes by use classification comprises the following:

Acetate rayon dyes: Developed for cellulose acetate and some synthetic fibers

Acid dyes: Used for coloring animal fibers via acidified solution (containing sulfuric acid, acetic acid, sodium sulfate, and surfactants) in combination with amphoteric protein

Azoic dyes: Contain the azo group (and formic acid, caustic soda, metallic compounds, and sodium nitrate), especially for application to cotton.

Basic dyes: Amino derivatives (and acetic acid and softening agents); used mainly for application on paper

Direct dyes: Azo dyes, and sodium salts, fixing agents, and metallic (chrome and copper) compounds; used generally on cotton-wool, or cotton-silk combinations

Mordant or chrome dyes: Metallic salt or lake formed directly on the fiber by the use of aluminum, chromium, or iron salts that cause precipitation in situ.

Lake or pigment dyes: Form insoluble compounds with aluminum, barium, or chromium on molybdenum salts; the precipitates are ground to form pigments used in paint and inks

Sulfur or sulfde dyes: Contain sulfur or are precipitated from sodium sulfide bath; furnish dull shades with good fastness to light, washing, and acids but susceptible to chlorine and light

Vat dyes: Impregnated into fiber under reducing conditions and reoxidized to an insoluble color.

Chemical classification is based on chromogen. For example, nitro dyes have the chromophore–NO

The Color Index (C.I.), published by the Society of Dyers and Colourists (United Kingdom) in cooperation with the American Association of Textile Chemists and Colorists (AATC), provides a detailed classification of commercial dyes and pigments by generic name and chemical constitution. This sourcebook also gives useful information on technical performance, physical properties, and application areas.

Dyes are synthesized in a reactor, filtered, dried, and blended with other additives to produce the final product. The synthesis step involves reactions such as sulfonation, halogenation, amination, diazotization, and coupling, followed by separation processes that may include distillation, precipitation, and crystallization. In general, organic compounds such as naphthalene are reacted with an acid or an alkali along with an intermediate (such as a nitrating or a sulfonating compound) and a solvent to form a dye mixture. The dye is then separated from the mixture and purified. On completion of the manufacture of actual color, finishing operations, including drying, grinding, and standardization, are performed; these are important for maintaining consistent product quality.

1.2   Manipulating the color and application of dyes

This section considers some of the chemistry behind the colour of dyes and how the target material, for example a fiber, influences the method of dyeing and the dye used.

A dye in solution is coloured because of the selective absorption of certain wavelengths of light by specific bonds in the molecule. The light that is transmitted is seen by the observer and appears coloured because some of the wavelengths of the visible spectrum are now missing.

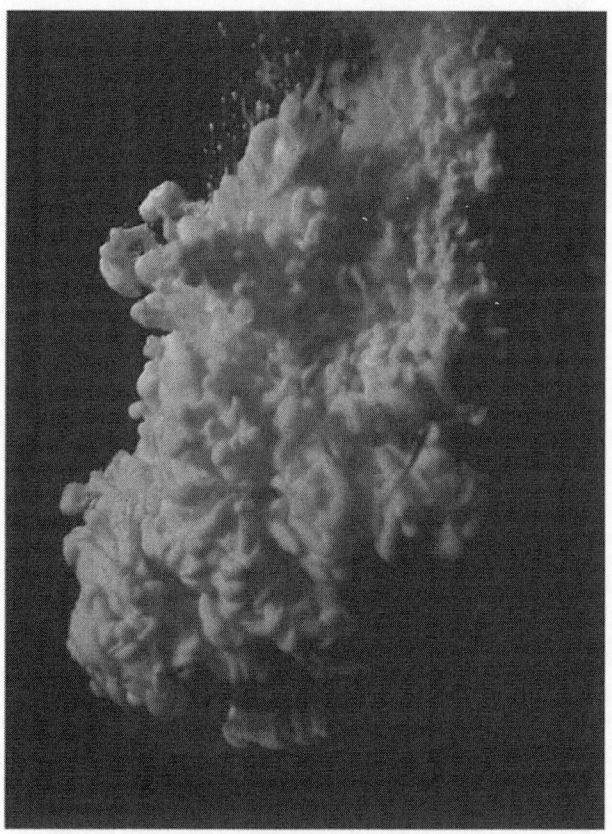

Figure 1.1

Mauveine was the first ever-synthetic dye. It was produced accidentally by William Perkin in 1856 who was trying to synthesize quinine. It became particularly popular when Queen Victoria wore a silk gown dyed with mauveine at the Royal Exhibition of 1862 in London [By kind permission of the Society of Dyers and Colourists].

The absorption of visible light energy by the compound promotes electrons in the molecule from a low energy state, the *ground state*, to a higher energy state, the excited state. The molecule is said to have undergone an electronic transition during this excitation process. Particular excitation energies correspond to particular wavelengths of visible light.

It is a pi (Π) electron (an electron in a double or triple bond) that is promoted to the excited state. Even less energy is required for this transition if alternate single and double bonds (i.e. conjugated double bonds) exist in the same molecule. The excitation of the electron is made even easier by the presence of aromatic rings because of the enhanced delocalization of the pi electrons.

By altering the structure of the compound, colour chemists can alter the wavelength of visible light absorbed and therefore the colour of the compound.

The molecules of most coloured organic compounds contain two parts:

(i) A single aryl (aromatic) ring such as benzene or a benzene ring with a substituent. Alternatively there may be a *fused ring system* such a naphthalene (two rings fused together) or anthracene (three rings fused together).

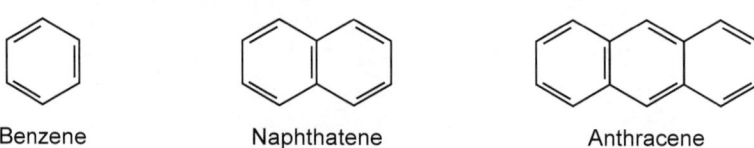

Benzene   Naphthatene   Anthracene

Figure 1.2

Where the rings join, they share two carbon atoms and thus naphthalene with two rings has 10 carbon atoms, not 12. Similarly, anthracene has 14 carbon atoms rather than 18. As naphthalene and anthracene contain delocalised electrons over *all the rings* it is inappropriate to use the delocalised symbol which is used for benzene in the other units, for that would indicate two or three separate delocalised systems. Thus in this unit, *Kekule structures* are used.

(ii) An extensive conjugated double bond system containing unsaturated groups, known as *chromophores*, such as:

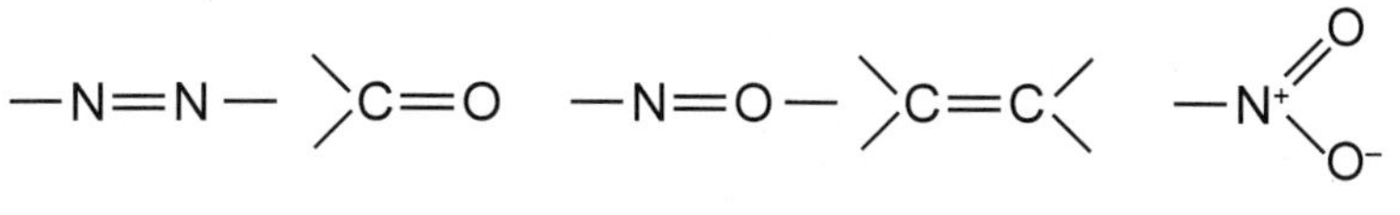

Figure 1.3

The intensity of colour can be increased in a dye molecule by addition of substituents containing lone pairs of electrons to the aryl ring such as:

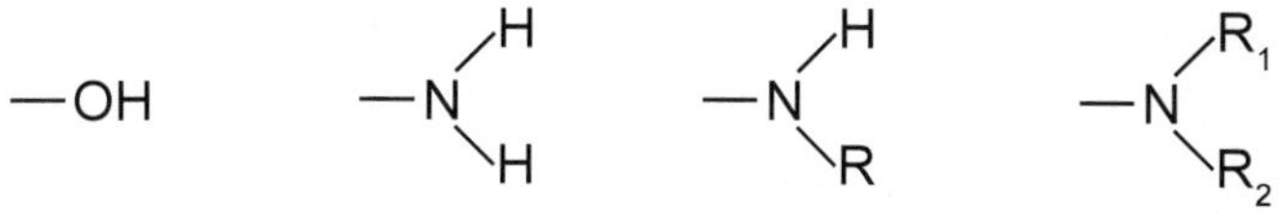

Figure 1.4

These groups are known as *auxochromes*.

Sometimes the entire structure of the colorant is called the *chromogen*.

To make the colorant of importance industrially, colour chemists must also be able to alter the compound's solubility, and groups may be included to make the colorant soluble in water. Examples include the sulfonic acid group, $-SO_3H$, or the carboxylic acid group, $-COOH$, or more usually, the sodium salt of these acids, $-SO_3^-Na^+$ and $-COO^-Na^+$, respectively. Another key concern of chemists developing dyes is to enhance its reactivity with the object that they want to colour, for example the molecules of the fiber. This is discussed below and examples are given throughout the unit.

1.2.1   The coloration of textiles

The chemical nature of a dye is determined by the chemical and physical properties of the fibers of the textile to be coloured. The four main types of fibers (Table 1.1) are protein, cellulosic, regenerated (based on cellulose or derivatives) and synthetic.

During the process of dyeing a textile, the dye is distributed between the two phases, the solid fiber phase and the aqueous phase, and at the end of the dyeing process the solution is depleted and most of the dye is associated with the fiber. Once the dye molecules penetrate the fiber there is immediate interaction between the two components, which prevents desorption of the dye molecules back into solution. The type of interaction, whether physical or chemical, will depend on the groups on the dye molecules and in the fiber chains (Table 1.2).

Table 1.1 Classification of textile fibers.

Natural fibers		Manmade fibers	
Protein	**Cellulose**	**Regenerated**	**Synthetic**
Wool	Cotton	Viscose rayon	Polyamides
Silk	Linen	Cellulose ethanoates	Polyesters
Mohair	Ramie		Acrylics
Cashmere			

The term regenerated is used when a natural polymer has been treated chemically to form another polymer. For example, natural cellulose from plants, when treated with ethanoic anhydride (acetic anhydride), produces a polymer, cellulose ethanoate, which is rayon.

Table 1.2 Approximate relative strengths of bonding between a dye and a fabric.

Bond type	Approximate relative strength
Covalent	30.0
Ionic	7.0
Hydrogen	3.0
Other intermolecular	1.0

Figure 1.5

Before a colorant is used, its light fastness must be determined. These racks (Fig. 1.5), situated on the Northeast coast of Australia, are used for many tests of weathering, amongst them being colourfastness. The position of the racks can be altered but in the photo, they are at an angle of 45° to the horizontal [By kind permission of the Allunga Exposure Laboratory].

The colorfastness of a colored textile is defined as its resistance to change when subjected to a particular set of conditions. The dye should not be affected greatly by sunlight (light fastness), heat when the fabric is ironed (heat fastness), perspiration (perspiration fastness) and when washed (wash fastness).

1.3   Classification of colorants

The *Colour Index International,* produced by the Society of Dyers and Colourists, in Bradford, is a comprehensive list of known commercial dyes and pigments and is updated regularly. Each colorant is given a Colour Index (C.I.) Name and Number. For example:

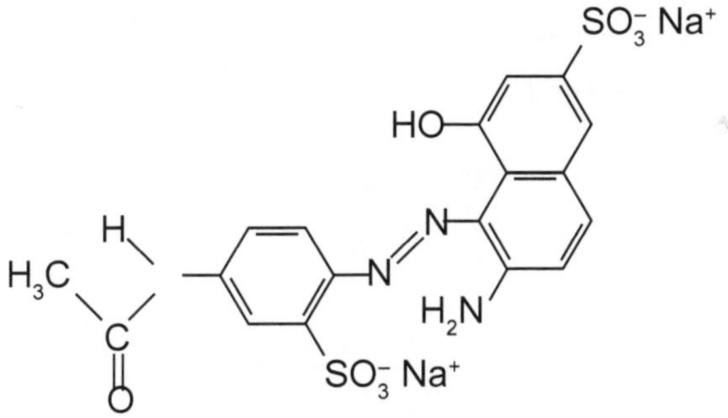

C.I. Acid Red 37

Figure 1.6

All the colorants in the list have been classified by their chemical structure and by their method of application.

1.3.1   Classification of colorants by their chemical structure

The Colour Index assigns dyes of known structure to one of 25 structural classes according to chemical type. Amongst the most important are mentioned below:

(a) Azo dyes
(b) Anthraquinone dyes
(c) Phthalocyanines

(a) Azo dyes

The azo dyes constitute the largest chemical class, containing at least 66% of all colorants. The characteristic feature is the presence in the structures of one or more azo groups, together with hydroxyl groups, amine and substituted amine groups as auxochromes.

$$-N=N-$$

Figure 1.7

Aromatic azo compounds are produced from aromatic amines via the corresponding diazonium salt.

A diazonium salt is formed when an aromatic amine is treated with nitrous (nitric(III)) acid. The nitrous acid is formed in situ by adding dilute hydrochloric acid to a cool solution of sodium nitrite at ca 278 K. In the following example, a solution of benzenediazonium chloride has been formed from phenylamine (aniline), the simplest aromatic amine:

$$NaNO_2(aq) + HC(aq) \longrightarrow HNO_2(aq) + NaCl(aq)$$

Figure 1.8

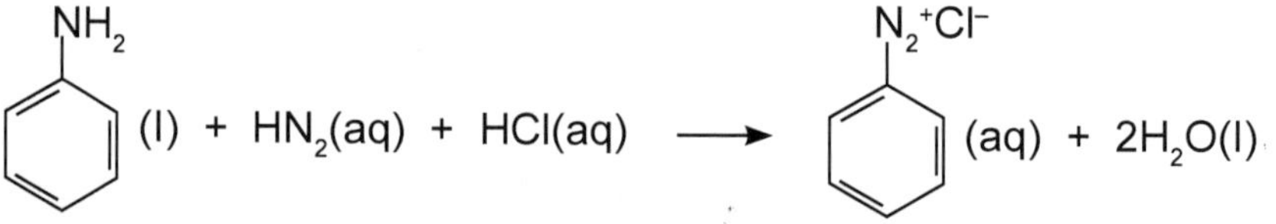

Aniltine   Benzenediazonium chloride

NH_2 (l) + HN_2(aq) + HCl(aq) $\longrightarrow$ $N_2^+Cl^-$ (aq) + $2H_2O$(l)

Figure 1.9

A solution of another compound such as another aromatic amine or a phenol is then added to the cool solution and produces an azo compound which is coloured. One example is the formation of a red dye when an aqueous solution of 4-aminonaphthalenesulfonic acid (naphthionic acid) is added to a solution of 4-nitrobenzenediazonium chloride to form C.I. Acid Red 74:

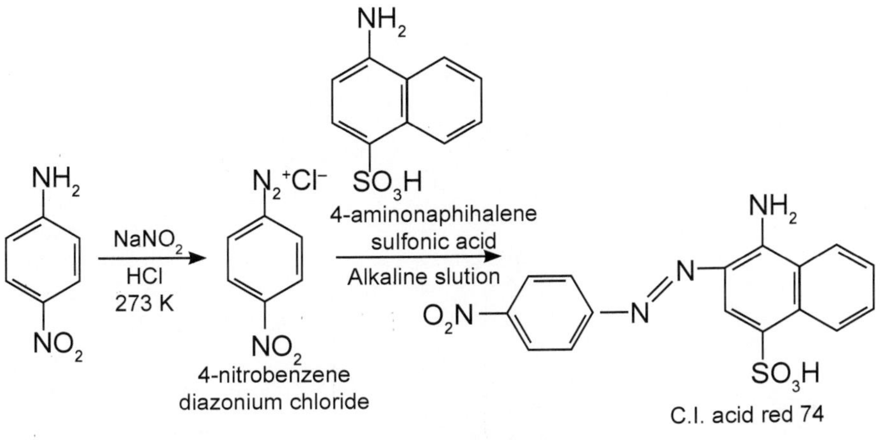

Figure 1.10

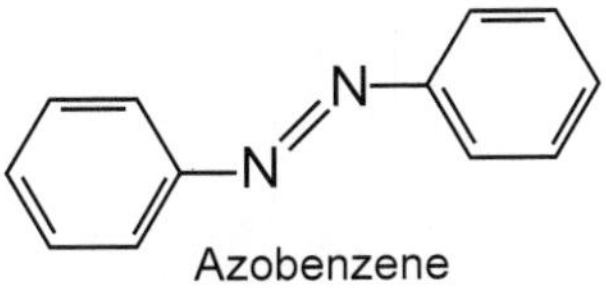

Azobenzene

Figure 1.11

Azobenzene is the chromophore of these azo dyes, and the colour of the molecule can be modified and the intensity of colour increased by varying the auxochromes (Table 1.3).

Table 1.3 The molecular structures of some azo dyes showing the auxochromes.

Structure	Colour observed
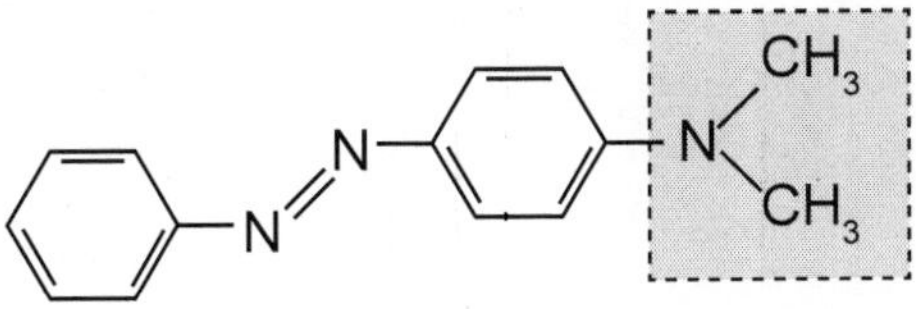 **Figure 1.12**	Yellow-green
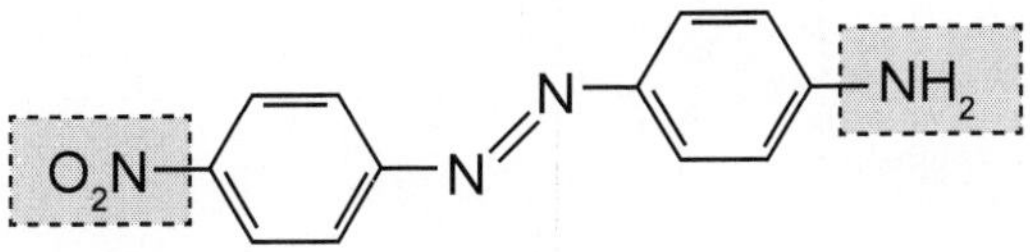 **Figure 1.13**	Yellow

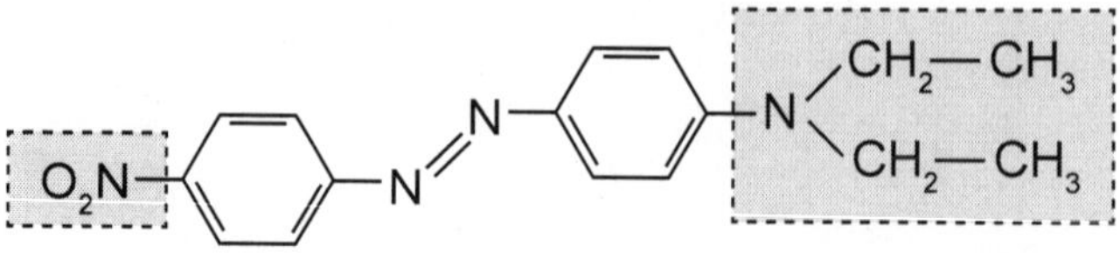

Figure 1.14

Red

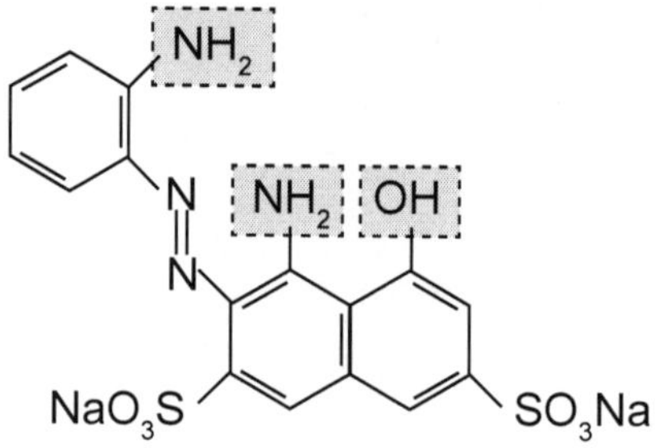

Blue

Figure 1.15

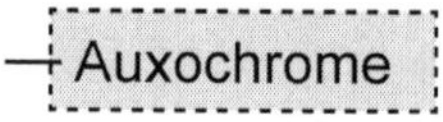

Figure 1.16

Some azo dyes, those containing a hydroxy group ortho (or para) to the azo group, for example, C.I. Acid Orange 7, exhibit tautomerism, a process in which the molecule exists as two or more different structures in equilibrium. The hydrogen atom on the hydroxyl group is able to migrate to the nitrogen atom of the azo group and vice versa:

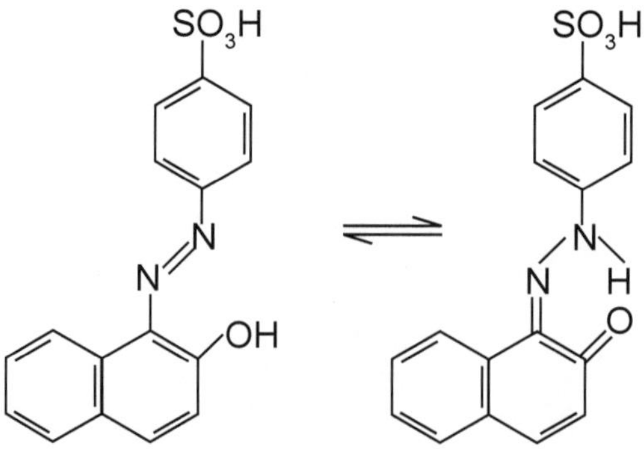

C.I. acid orange 7

Figure 1.17

This type of tautomerism involves equilibrium between a hydroxyazo tautomer and a ketohydrazone tautomer, although the ketohydrazone tautomer generally dominates and the colour observed is of longer wavelength (a bathochromic shift).

(b) Anthraquinone dyes

Anthraquinone dyes account for about 15% of colorants and have structures based on quinones. The simplest quinone is benzoquinone, which has two isomers:

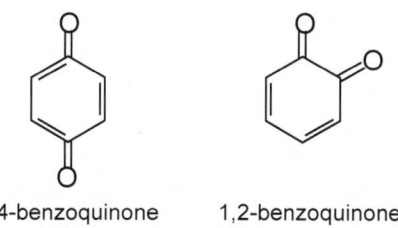

1,4-benzoquinone   1,2-benzoquinone

Figure 1.18

Anthraquinone, the simplest of the anthraquinones, is based on anthracene:

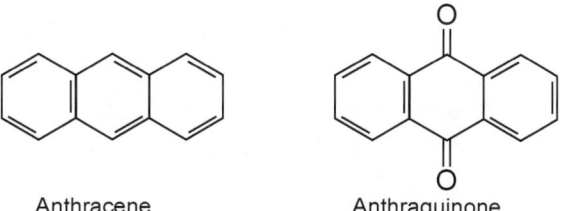

Anthracene   Anthraquinone

Figure 1.19

Two well-known examples of anthraquinones which are used as dyes are C.I. Disperse Red 60 and C.I. Disperse Blue 60:

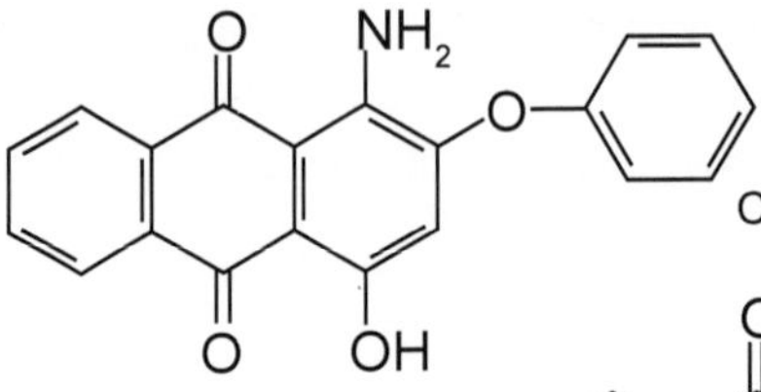

C.I. Disperse red 60

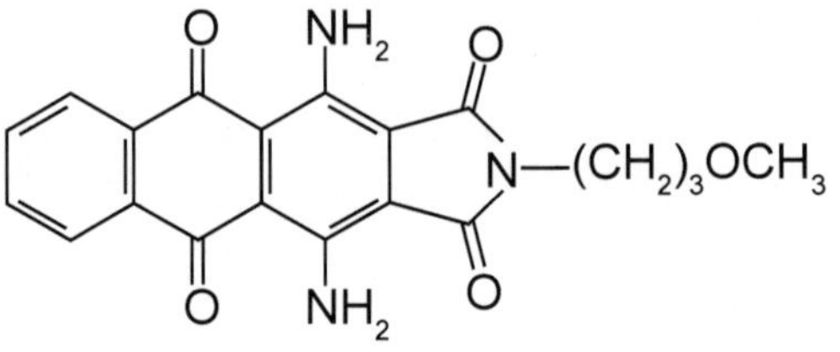

C.I. Disperse blue 60

N—(CH$_2$)$_3$OCH$_3$

Figure 1.20

(c) Phthalocyanines

Phthalocyanines are essentially made up of four molecules of isoindole:

These molecules are connected to each other in a phthalocyanine by nitrogen atoms. The structure of phthalocyanine is:

N—H

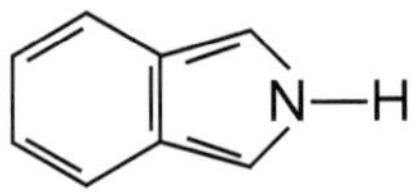

Figure 1.21

These molecules are connected to each other in a phthalocyanine by nitrogen atoms. The structure of phthalocyanine is:

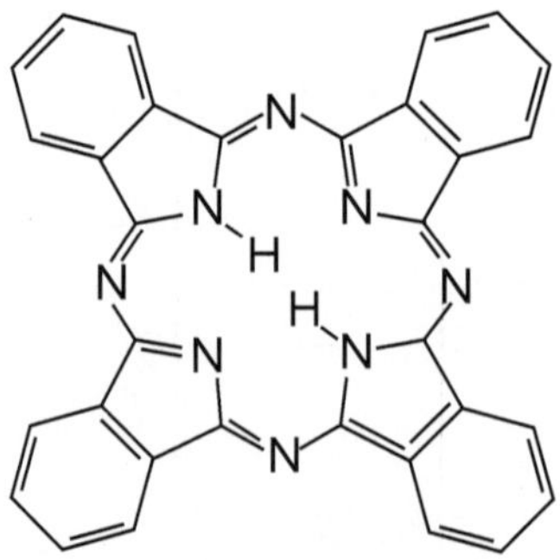

Figure 1.22

Phthalocyanines coordinate with metal atoms. The most important, contributing about 2% of all colorants, are the copper phthalocyanines, used for their brilliant blue and green colors. An example is C.I. Direct Blue 86:

C.I. Direct blue 66

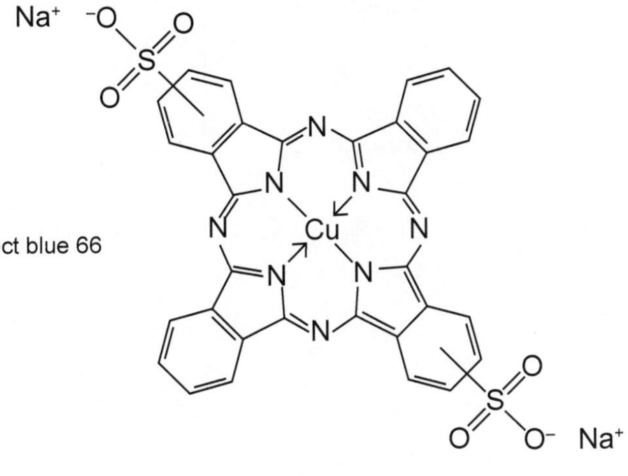

Figure 1.23

The sulfonic acid groups assist the solubility of the dye in water.

The formula indicates that the sulfonic acid groups can be at different positions on the aromatic rings.

1.3.2   Classification of colorants by methods of application

Classification by the method of application is important to the textile dyer applying the dye to produce the colour required. To obtain the required shade, the dyer usually has to make mixtures of dyes and must ensure that these are compatible.

The basic features that control dye transfer from solution to fiber are:

- the pH of the solution in the dyebath (for acid and basic dyes)
- an electrolyte (a solution of sodium sulfate or chloride)
- the temperature (within the range of ambient to 400 K)
- chemicals, known as dispersing agents, that produce a stable aqueous dispersion of dyes of very low solubility

Table 1.4 lists the dyes under their technological names that indicate how they are applied, along with the fibers to which they are applied.

Table 1.4 Technological classification of dyes.

Dye	Fiber
Group 1	
Acid	Wool and other protein fibers, polyamides
Metal-complex	Wool and other protein fibers, polyamides
Direct	Cotton, linen, viscose
Basic	Acrylic
Disperse	Polyesters, polyamides, ethanoates
Group 2	
Reactive	Cotton, linen, viscose, wool, silk
Vat	Cotton, linen, viscose
Sulfur	Cotton, linen

2
Direct dyes

2.1   Introduction

Although Perkin synthesised the first synthetic dye, Mauve, thirty years earlier, the discovery of first fibre-substantive direct dye, Congo Red in 1884 was made by Bottinger, actually initiated the period of major growth and establishment of the synthetic dyestuff industry. The name direct cotton colour (DCC) or simply, direct dye, was derived from the fact that it was the first synthetic dye, which had inherent substantivity for cotton. Earlier, cotton usually had to be mordanted with either natural mordants or their synthetic replacements, before application of dyes. The most attractive feature of direct dyes is the essential simplicity of the dyeing process and lower cost for achieving high depth of shades.

At present a great number of direct dyes are known which give a full range of colours and find wide use to cellulosic textiles, natural proteins, paper and leather. The chromophore in direct dyes includes azo (disazo, trisazo and polyazo dyes), stillbene, oxazine and phthalocyanine, with some thiazole and copper complex azo dyes. Dyes containing cyclic imide system, e.g. imidazolone, quinoxilinedione or benzoaxazolone show significant affinity to cellulose, even in cases where the dye has a relatively small molecular size. Direct dyes contain one or more sulphonic acid groups to impart solubility. Direct dyes are also called substantative colors because cellulosic fibres have strong affinity for them. Solubility decreases with increase in molecular weight of dye. While a major portion of the production of direct dyes goes for coloration of paper, leather and other uses. Direct dyes are water-soluble and can be applied to cotton and other cellulosic fibres such as viscose, linen, jute, hemp and ramie.

Direct dyes have varying fastness to washing, light, perspiration and other wet fastness properties, and varying staining properties on wool, silk, polyester and acrylic fibres. Most direct dyes have limited wet fastness in medium to full shades unless they are after-treated, but some are better than others. Resin finishing after dyeing produces a notable improvement in wet fastness, especially on regenerated cellulosic fibres. The light fastness varies from very poor (rating 1) to outstanding (rating 8).

Direct dyes are used in low-priced viscose or blended curtain fabrics, furnishings and carpets. Cheap cotton apparels, casual wear and bedspreads, low quality discharge-printed materials, which are not washed frequently, are dyed with direct dyes.

Direct dyes are sold under various brand names such as Solar, Pyrazol (Clariant), Incomine (Indochem), Sirius, Sirius Supra (Bayer), Chlorantine, Cuprantine, Solophenyl (Ciba), etc.

Nearly all direct dyes are azo dyes containing one or more sulphonic acid (as sodium salt) groups to impact water solubility. Most direct dyes contain two or three azo groups. Direct dyes of the ordinary type comprise a sizeable portion of the total synthetic dyes. They are easy to apply and cheap. Direct dyes are made by diazotizing an aromatic amine with or without a sulphonic acid group. Cotton is mainly dyed by direct dyes because it is the cheapest among all classes of dyes. The relative ease and economy of preparation of azo dyes by the classic diazotization and azo coupling reactions have led to the synthesis of many dyes. The use of heterocyclic diazo and coupling components has made the production of brighter colors possible.

The fastness of azo dyes varies widely with structure and environment, but in general it tends to be less than that of anthraquinone vat dyes and indigoid dyes and greater than that of sulphur dyes.

Direct dyes are particularly useful for dyeing fibres made from cellulose.

About 70% of all direct dyes are unmetallised azo compounds. Congo Red is synthesized from Benzidine which is universally recognized to be a carcinogen. Direct dyes were first applied to paper and leather because of their poor wash and light fastness properties. Bayer replaced benzidine by 0-tolidine to obtain a dye of higher stability and substantivity. Benzopurpurin 4B has wash fastness property. The disazo stilbene Chrysophenine G with good all-round fastness soon became the foremost yellow. Diamine Green B, a trisazo structure, was made from benzidine and other cheap intermediates. In 1915, Ciba introduced certain dihydroxy azo dyes with well-defined 1:1 copper complex and marketed as Chlorantine fast dyes. The significant developments in the field of direct dyes was the introduction of Indosol SF dyes by Sandoz in 1981 contained twelve members copper complex dyes which require after treatment to improve the light fastness. Ciba in 1992 introduced the Solofix system that was somewhat analogous to the Indosol approach. In 1995, Sandoz introduced Optisal range of metal-free direct dyes, which is suitable for use with disperse dyes.

Direct dyes are mainly used for cotton and rayon and can be applied to silk, jute and wool. They are, for the most part, derived from benzidine and their molecule contains two azo groups N=N; they are, therefore, also known as 'diamine colours'. The direct dyestuffs are marketed under various trade names such as Atul Direct (Atul), Chlorazol and Durazol (ICI), Chlorantine Fast and Direct Fast, Cuprantine, Solophenyl (Ciba), Sirius and Sirius Supra (Bayer), Incomine and Incomine Light Fast (Indokem), Solar, Pyrazol (Clariant) Rano, Ranofast and Rano Sunfast (SuhridGeigy). Some of these dyes are discontinued now. But older dyeing technicians have used long back.

The fastness properties of some of the dyes can be improved by an after treatment with metallic salts; generally copper salts improve fastness to light and chromium salts, fastness to washing.

While a major portion of the production of direct dyes goes to the textile industry, approximately 25% of the production of the dyes goes for coloration of paper, leather and other uses. Direct dyes are water-soluble and can be applied to cotton and other cellulosic fibres such as viscose, linen, jute, hemp and ramie.

Direct dyes have varying fastness to washing, light, perspiration and other wet fastness properties and varying staining properties on wool, silk, polyester and acrylic fibres. Most direct dyes have limited wet fastness in medium to dark shades unless they are after-treated, but some are better than others. Resin finishing after dyeing produces a notable improvement in wet fastness, especially on regenerated cellulosic fibres. The light fastness varies from very poor (rating 1) to outstanding (rating 8). Direct dyes are used in low-priced viscose or blended curtain fabrics, furnishings and carpets. Cheap cotton apparels, casual wear and bedspreads, low-quality discharge-printed materials, which are not washed frequently, are dyed with direct dyes.

At present a great number of direct dyes are known which give a full range of colours and find wide use to cellulosic textiles, natural proteins, paper and leather. The chromophore in direct dyes includes the following:

(a) Azo (disazo, trisazo and polyazo)

(b) Stilbene

(c) Oxazine

(d) Phthalocynanine

Besides the above four groups there are some thiazole and copper complex azo dyes. Dyes containing cyclic imide system, e.g. imidazolone, quinoxilinedione or benzoxazolone show significant affinity to cellulose, even in cases where the dye has a relatively small molecular size. Direct dyes contain one or more sulphonic acid groups to impart solubility. They are in general duller than the fibre reactive dyes. Goods dyed with direct dyes unless given a proper after-treatment tend to bleed with every wash. The direct dyes in many cases exhibit a better light fastness as compared to the reactive dyes.

A detailed study by an SDC (Society of Dyers and Colourists, U.K.) committee showed that the four important parameters in defining the dyeing properties and compatibility of direct dyes are:

1. Migration or leveling power
2. Salt controllability
3. Influence of temperature on exhaustion
4. Influence of liquor ratio on exhaustion

2.2   Classification of direct dyes

As in other classes of dyes, the direct dyes can also be classified in many ways like the chemical nature of the chromophore, or by their dyeing characteristics and performance properties, but in general the most popular way of classification by the dyes is by means of their dyeing characteristics.

The classification of direct dyes by SDC is based essentially on the compatibility of different groups of direct dyes with one another under certain conditions of batch dyeing. They are classified into three categories, such as:

1. *Class A* – Dyes belonging to this group migrate well and therefore have high leveling power. These dyes are also called self-leveling dyes. When dyeing these dyes, they may dye unevenly in the initial period but continued dyeing levels out the shade. These groups of dyes do not need the addition of salt for exhausting.

Typical examples of these groups are C.I. Direct Yellow 50, C.I. Direct Red 31 and C.I. Direct Blue 67 and maximum exhaustion is reached at 60–80°C.

2. *Class B* – These dyes have poor leveling properties and are not self-leveling dyes. Their exhaustion has to be brought about by controlled addition of salt. These are called salt-controllable dyes. The standard dyes of this group are C.I. Direct Red 26, C.I. Direct Blue 8 and C.I. Direct Violet 1. Maximum exhaustion is obtained at 80–100°C in presence of 5 g/l salt and when dyeing is carried without salt, the exhaustion is markedly inferior.

3. *Class C* – These dyes are poor leveling colours and are highly sensitive to the presence of salt. The exhaustion of these dyes cannot adequately be controlled by the addition of salt alone and they require additional control by the proper rate of raising the temperature of the dyebath. These dyes reach equilibrium at temperatures higher than 100°C. These are described as temperature-controllable dyes.

Thus, knowledge of the group into which a dye falls is a valuable guide to satisfactory dyeing when more than one dye is required to match a given shade. Thus, a salt-sensitive dye should not be mixed with a dye which requires a large amount of salt to effect exhaustion. Also while choosing dyes of the same group, their rate of exhaustion must be approximately be the same.

2.3   Application of direct dyes

The direct dyes are used for the dyeing of various cellulosic materials. They can be dyed in fibre, yarn, fabrics and garment forms.

They can be used in the following processes:

1. Exhaust process
2. Semi-continuous process
3. Continuous process

The exhaust process includes dyeing of cotton yarn in hank form in cabinet hank dyeing machines or cotton yarn in package form in HTHP dyeing machines. Piece goods are dyed in winch machine or jigger. The dyeing is usually commenced at lower temperature and lower salt content and the temperature gradually raised to boil and the quantity of salt increased. Salt is also added in portions. In dyeing light shades, Dispersol VL or Fibrolev DLV is added in small quantities to get level dyeings.

The best temperature for dyeing the yarn varies with different dyes. In using mixtures of dyes for compound shades, only those dyes which possess approximately same dyeing qualities are chosen.

The semi-continuous process includes the following:

1. *Pad-Jig* – The fabric is padded with direct dye and then fixation of the dye is done in jigger.

2. *Pad-roll* – It is the most suitable semi-continuous method which is followed by:

(a) *Cold Pad batch* – The fabric after padding with direct dye is wrapped in a cylinder and the fabric roll is rotated slowly at room temperature for 8–24 hours depending on the depth of the shade.

(b) *Hot Pad batch* – The fabric is first padded with direct dye at 50–80°C and then passed through an infra-red heater (80–90°C) before it is wound on a rotating cylinder.

Continuous process is much less suitable using direct dyes as compared to batch-wise dyeing. Tailing is a serious problem in continuous dyeing. Following are the process being followed:

1. *Pad-steam* – The fabric is padded with direct dye and urea along with 1–3 gpl sodium pyrophosphate. Urea increases the solubility of the dye and eases its diffusion during steaming.

Sodium pyrophosphate acts as a buffer. The padded fabric is then steamed at 100–105°C for 30 seconds to 5 minutes. Sodium alginate may be added to the pad bath to avoid migration of the dye due to the humidity of the steam.

2. *Pad-dry* – The fabric is first padded with direct dye and then dried in a suitable machine to fix the dye. Rinsing, subsequent after-treatments and final rinsing are carried out as usual.

3. *Pad-salt* – The fabric after padding with direct dye is immediately passed through hot saline bath containing 10–20 gpl salt.

Many direct dyes are suitable for combined scouring and dyeing process employing soda ash and a non-ionic detergent. Dyes with amide groups should be avoided.

Combined peroxide bleaching and dyeing can be carried out with selected direct dyes. The materials should be thoroughly prepared. Soda ash is preferred to caustic soda and organic stabiliser to sodium silicate.

2.4   After-treatment of direct dyed material

The lack of wet-fastness of direct dyes may be undesirable for the materials, which are expected to withstand washing. Many methods are used to increase the molecular weight of the dye, after it has been adsorbed by the fibre, to render less soluble in water and, therefore, more fast to wet treatments. The methods are described below:

1. *Diazotisation and Coupling* – Direct dyes containing amino groups are applied in the normal way, washed off and treated in a bath containing sodium nitrate and hydrochloric acid, suitably buffered. The material is then rinsed and passed into a bath containing dissolved coupling component. Such compounds, usually referred to as developers, are phenolic compounds like resorcinol, m-phenylene diamine and β-naphthol and are dissolved in sodium hydroxide. The treatment in developer solution is carried out for 15–20 minutes and then washed off. A few direct dyes which can be diazotized and developed are:

- C.I. Direct Yellow 64 and 65
- C.I. Direct Reds 127, 136 and 155
- C.I. Direct Blues 2 and 20
- C.I. Direct Green 52

The improvement in fastness rating may be of the order of 1–2 units.

2. *Coupling with Diazotized Amine* – The cellulosic material is dyed with a direct dye containing an amino or hydroxyl group (C.I. Direct Brown 152) and then the dyed material is treated with a solution of diazotized base, usually paranitroaniline. Virtually a new dye is produced in situ in the fibre.

3. *Treatment with formaldehyde* – The direct dyed material is rinsed and run in a liquor containing 2–3% (owm) formaldehyde (40% solution) and 1% acetic acid (30%) at 70–80°C for 30 minutes. The improvement in fastness rating is of the order of 1.

4. *Treatment with metal salts* – For improving fastness to light as well as washing the direct dyed material is given a treatment with the following:

Potassium dichromate – 1–2% (owm)

Copper sulphate – 2–4% (owm)

Acetic Acid – (30%) – 1–5%

Heated for 30 minutes at 80°C and then rinsed in cold water. Copper sulphate is used for increasing fastness to light and bichromate for increasing fastness to washing.

5. *Treatment with cationic fixing agents or synthetic resins* – In order to improve the brightness or brilliancy of shade, direct dyeings are sometimes topped with a small quantity of basic dye in a separate bath. The direct dye acts as a mordant for the basic dye so that the basic dye becomes fixed on the material producing a compound shade. Usually while topping, a basic dye of the same general colour as the direct dye is used. Thus, material dyed with Direct Green is topped with a small quantity (0.1%) of Malachite Green to give a brilliant green shade.

Now-a-days direct dyed material is treated with cationic fixing agents making a greater molecule and less water solubility and better wash fastness. However, most of these compounds tend to lower light fastness of the dyed materials and alter the shade. A number of synthetic resin compounds are also used.

Though these dyes have out-dated now due to toxic nature, bad eco-friendly behaviour and colourfastness properties not as per standard etc. But new developed direct dyes have introduced nontoxic, eco-friendly and moderate colourfastness rating, etc. Though direct dyes cannot compete the merit of reactive dyes in view of colour fastness to washing, wet crocking, dry crocking, sea water, perspiration and light fastness, etc., which is prime drawback of such dyes. These dyes are used due to its economical costing.

2.5   Chemical structure of direct dyes

The relationship between the chemical structure of a direct dye and its substantivity for cellulose is an intriguingly complex one. In general, the affinity of direct dyes can be increased by enlarging the planar, conjugated double-bond system. However, the forces of attraction between the direct dye and cellulosic fibres may include hydrogen bonding, dipolar forces and non-specific hydrophobic interactions which are highly dependent upon the nature of the dye structure and the polarity of the dye molecule.

They are synthesized with sulfonic acid groups to give them solubility in water, dissociating to give sodium cations and the anionic dye species. They are also designed so that they are as linear and planar in structure as possible. This allows the dye to be attached to the cellulosic chains in the fibre, often via intermolecular (including hydrogen) bonding.

They are applied in the dyebath in aqueous solution which contains sodium chloride. The salt reduces the electrical forces of repulsion between the negative charge on the fibre surface and the anionic dye species.

Azo dyes constitute the major proportion of direct dyes. They are made by diazotizing an aromatic amine and then coupling with an aromatic compound containing a hydroxyl, amino or other group which creates a coupling position. Polyazo dyes are made either from a diamine like benzidine, o-tolidine, o-dianisidine or by diazotizing an amine and coupling with another amino-compound, the amino group of which may then be diazotized and further coupled with suitable compounds.

The thiazole type of direct dyes does not cover the entire range of color and are mainly restricted to yellow, orange or brown. They are characterized by the presence of thiazole ring and form an azoic compound when amine groups are present.

At least 70% of all direct dyes listed in Colour Index are unmetallised azo compounds. In contrast to other dye-classes, the great majority of them are disazo (about 50%) or polyazo (about 33%) types, the former predominating in the brighter yellow to blue sector and the latter in the duller greens, browns, greys and blacks. The copper complex direct dyes derived from the above two types are mostly duller violets, navy blues and blacks. Stilbene and thiazole direct dyes, the non-azo chromogens in the yellow to red sector, also bear certain similarities to the above two types. The phthalocyanine and, more recently, dioxazine chromogens are utilized for producing bright blue direct dyes of high light fastness.

An important group of direct dyes is derived from stilbene such as C.I. Direct Yellow 19. Some important yellow, orange and brown direct dyes are based on stilbene. Primuline is the first of the thiazole direct dye.

Most of the direct dyes are azo compounds, often containing two or three azo groups. Examples include C.I. Direct Orange 25 which has –OH, –NHCO– and –N=N– groups all of which have the potential to form hydrogen bonds with the hydroxyl groups in cellulose:

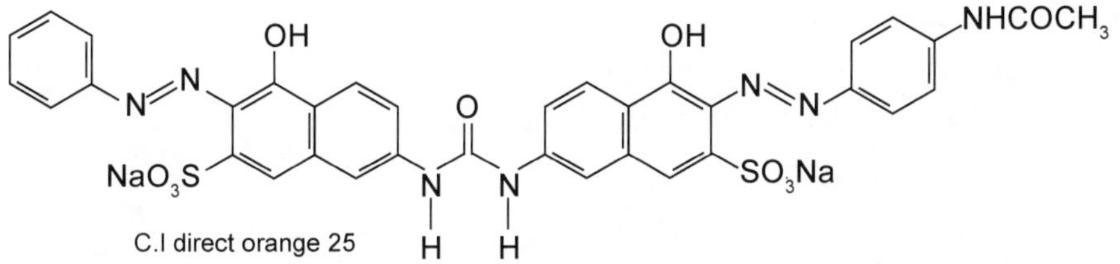

Figure 2.1

The colorant exhibits tautomerism, as there are two hydroxyl groups ortho to the azo groups. One of the tautomers in equilibrium with this form is where there are two ketohydrazone groups.

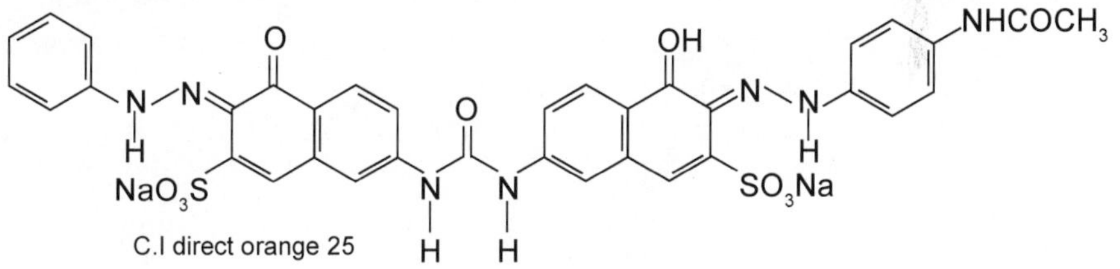

Figure 2.2

Another example, C.I. Direct Blue 71, has three azo groups, one of which is present as the ketohydrazone tautomer:

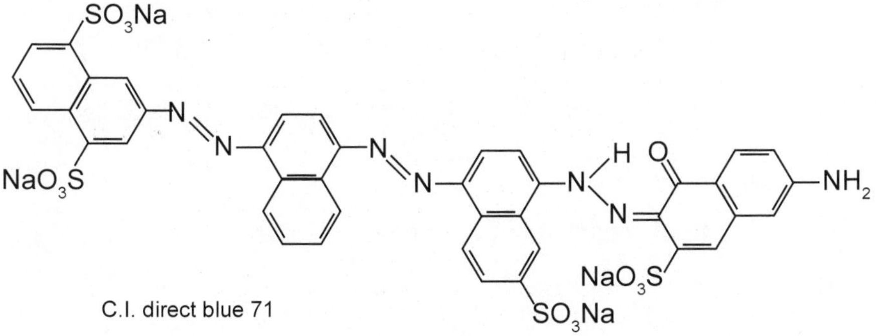

Figure 3.3

2.6   Manufacturing of direct dyes

1. *Sky Blue FF (Direct Blue 1)* – The raw materials required for producing 100 kg of direct dyestuff are given below:

- O-Dianisidine – 25 kg
- Chikago Acid (acid sodium salt) 75 kg
- Hydrochloric Acid (conc.) – 45 kg
- Sodium Nitrite – 12 kg
- Sodium Carbonate (Soda Ash) – 70 kg

Following are the steps:

(a) *Diazotization* – In a reactor, 25 kg of Dianisidine is dissolved in 15 kg of HCL acid and about 400 litres of boiling water. Then bring down the temperature by adding crushing ice piece to 5°C temperature with the help of a cooling jacket surrounding the reactor. Now remaining amount of HCL acid is added and then the Sodium Nitrite solution is added slowly in about an hour or so. Diazotization takes place and a clear Tetrazo solution is produced. After complete addition of sodium nitrite solution, check for excess of nitrite by starch iodide paper and confirm diazotization is complete.

(b) *Coupling* – Solution of Chikago Acid is prepared in minimum water, first by adding soda ash to make the ph of acid to neutral and then rest Soda Ash is added as 20% solution. When the solution is clear, bring down the temperature of chikago acid solution by adding ice to 0–5°C.

This solution is poured into the above Tetrazo solution slowly in about 1½ hours time and as the tetrazo is added go on adding ice and salt to bring down the temperature of the coupling mass to about 0–4°C. After stirring for a few hours, test for the completion of coupling and stir and mass overnight.

(c) *Isolation* – Next day salt out (5% common salt on volume basis) stir for 1 hour, filter, neutralise and dry.

2.7   Identification of direct dyed material

Small specimen (about 3 × 3 cm) of dyed fabric is successively treated with the following solutions at boil for 3–4 minutes with intermediate washing with water and squeezing.

(a) 50% dimethylformamide
(b) Concentrated dimethylformamide
(c) A mixture of glacial acetic acid and rectified spirit (1:1 v/v)

If complete stripping of dye is observed, presence of direct dye is confirmed.

To further confirm, boil fresh test specimen in 1% ammonium hydroxide solution for 1–2 minutes. The colour bleeds and the solution is distinctly coloured. Remove the test specimen and add few pieces of white bleached cotton, 25 mg of common salt to the above solution; boil for 2 minutes. Cool and rinse the cotton sample. The bleached cotton is dyed to approximately original shade. This confirms the presence of direct dye.

2.8   Identification of direct dye powder

Dissolve 150 mg of dye powder and 10 gm Glauber salt in 200 ml cold distilled water. Add 5 g of cotton fibre/yarn/fabric and dye the material by slowly increasing the temperature. After 5 minutes add 0.5 g Soda Ash and boil for about 10 minutes. Rinse the material and then boil with synthetic detergent for 2 minutes. By this, cotton gets dyed with only Reactive and Direct dyes. Dry the dyed cotton and treat with dimethyl formamide. If the colour is stripped, the powder is direct dye otherwise it is reactive dye.

2.9   Stripping of direct dye from the dyed fabric

In order to correct the shade, sometimes it may be necessary to strip the colour from the dyed material. Unless they have been after-treated, the stripping of direct dyes in most cases is easy. The colour can be stripped by boiling with Sodium Hydrosulphite or by bleaching with sodium hypochlorite solution having 1 to 3 grams per litre of available chlorine or bleaching with sodium chloride acidified with acetic acid or formic acid to pH 3–4. Among the three methods, stripping with sodium hydrosulphite is recommended, as it does not affect fabric strength.

In case the garment is treated with cationic fixing agent, the colour can be stripped only after removing the fixing agent. This can be done by boiling the garments with formic acid (2% owg) for 30 minutes. In case the fixing agent is not removed, it will act as mordant while re-dyeing the garments resulting in uneven dyeing.

The stripping recipe for direct dyes is as follows:

- 1 g/L Sodium carbonate
- 2 g/L Sodium hydrosulfite
- 2 g/L DA-BS800

Keep for 30–60 min at 50 °C, then rinse thoroughly.

Azoic dyes (Naphthol dyes)

3.1 Introduction

The discovery of diazonium compounds by Peter Griess in 1858 was exploited later on to develop a colored insoluble azo compound. In 1876, he demonstrated that azo dyes were readily obtained by coupling diazotized sulphanilic acid with various bases derived from benzene. However, insoluble azoic dyes were introduced in 1880 by Reed Holiday and generally referred to as ice colors or azoic colors. The first azoic dye was obtained from 2-naphthol and diazotized 4-nitroaniline, known as Para Red. Betanaphthol became a very popular component in the dyeing of azoic dyes. It was therefore a most important discovery in 1911 by the German Griesheim Elektron Company that β-naphthol could be replaced by Naphthol AS and the insoluble azoic color was developed. The Naphthol AS compounds were marketed under different trade names. The growth in this area was impeded by the outbreak of the World War in 1914. However, the number of color forming bases and their salts grew rapidly after the World War and a series of naphthols are now available with entire gamut.

Within the last decades interest in this class of colorants has markedly declined because of some concerns over the ecotoxicology of some of the azoic diazo components which are derived from aromatic amines, some of which are considered to be carcinogenic. A number of diazo components have been withdrawn from the market for this reason. During the period in 1975 to 1985, reactive azoic coupling components have been introduced. While azo dyes are readymade dyes, azoic colors are formed inside the textile materials by reaction of two colorless or faintly colored compounds called naphthols, or C.I. Azoic Coupling Components and fast bases, or C.I. Azoic Diazo Components. At first, the naphthols are applied on textile materials by a process called naphtholation as they have some affinity for the cellulosic materials. The treated material is subsequently treated with a soluble diazotized form of the base, when intense color is formed.

The latter step is known as coupling or development. The solublisation of the base is carried out with sodium nitrite and hydrochloric acid at low temperature and the process is diazotization.

Azoic colors give wide range from yellows, orange, scarlet, red, blues, blacks, but there are only few greens. Also called Naphthol Dyeing, is recommended to get the bright and fast shades in maximum depth similar to vat dyeing. Lemon yellow, dark orange, blood red, pink, red -orange, red brown, yellow brown, tan, dark red, violet, brown violet, coffee-violet, bright blue, china blue, navy blue, blue-black and royal blue are dyed successfully under this style. Except lemon, gold orange, brown, scarlet, red, blue, violet and black, rest colors have problems of dry and wet rubbing fastness; though this defect is removed by using efficient soaping and washing after dyeing. Few shades which are difficult to dye in vat dyes (due to no dyes available) are successfully dyed in this style. These are orange, scarlet, green and yellow colors.

The light and chlorine fastness depends on the combination used and can be as high as 7 and 5, respectively. Naphthol dyeing is employed for cheap quality fabric like khaddar, poplin, twill and net fabric. Blended fabrics are not recommended.

3.2   Stages of Naphthol dyeing

The Naphthol dyeing process consists of four stages:
- (i)   Impregnation of material with solution of suitable naphthols
- (ii) Removal of excess naphthols by squeezing or padding
- (iii) Development of color in solution of selected diazotized base
- (iv) Soaping

Following steps are followed:

(a) *Dissolution of Naphthols* – As such naphthols are insoluble in water so to make it in soluble form naphthols are pasted with any suitable wetting agent such as TRO. Then warm water is added under stirring. After making a uniform solution of naphthols, caustic soda flakes (pre-dissolved) are added slowly under stirring until clear solution obtained. Then formaldehyde is added in order to improve the stability of naphthol to air. The quantity of caustic soda flakes depends on the type of naphthols, selection of base and quantity of naphthols.

(b) *Preparation of Naphthol Bath* – The amount of naphthols depends on the shade required as below:
- Light shade – 1–5 gpl
- Medium shade – 5–10 gpl
- Heavy dark shade – 10–20 gpl

The naphthol bath is prepared with caustic soda, protective colloid and common salt. Caustic soda in the naphthol bath contributes to the stability of naphthol solution during treatment. Common salt is added to the bath to increase the affinity of naphthols when dyeing is carried out by exhaust method. The alkaline coupling component solutions have a tendency to become turbid on standing. This effect is minimized by the addition of protective colloids such as glue or gum arabic or wetting or dispersing agents. It is advisable to add these products together with a sequestering agent when hard water is used.

(c) *Impregnation of Naphthol Solution* – It is carried out by two methods:

(i)   Exhaust method

(ii) Padding method

In exhaust dyeing method jigger, winch, cabinet and warp reel machines are used successfully. A desired M: L ratio of 1:20, 1:25 is employed to get uniform results.

Cotton is mostly dyed in hank, warp or piece form. Hanks may be dyed in either in open beck, liquor circulating or in package form. For the application of piece goods, a pad mangle with double nip is the most favored method, although Jigger can be used if followed by an efficient squeeze on the mangles. For heavy cloths and for small production, it is advisable to use jigger. Jigger is not economical for naphthols of low substantivity. Fabric or yarn is impregnated with naphthol solution for about 20 minutes. While padding dyeing method is more economical than that of exhaust methods, fabric is dyed by continuous method in the pad-bath.

The time of treatment and temperature of the impregnation bath depends on the brand of naphthols.

After impregnation naphthol present in the fiber is partly mechanically held and partly substantively held. The mechanically held naphthol is generally removed by centrifuging in the case of yarn and by padding mangle in the case of fabric as the rubbing fastness depends to a large degree on this operation.

(d) *Base Application Process* – The second component for the production of insoluble pigments within the fiber consists in coupling with diazonium salt which can be coupled with naphthol already deposited on the fiber. The base is insoluble in water. To make it in soluble form, sodium nitrite and hydrochloric acid are used. This soluble form of base is used to get the desired shade on naphtholated material.

For example, Naphthol AS + Diazotised Fast Red TR Base = Bright Red Azoic Pigment.

The free bases are mostly derivatives of primary amines and are insoluble in water. To make it into soluble form, the bases are first converted into hydrochloride of base (IV) by reacting with concentrated solution of hydrochloric acid. Now add ice to maintain 10–15°C temperature. Then sodium nitrite is added slowly under stirring. Now keep this solution for 10–15 minutes for uniform reaction.

Now filter the prepared base to transfer into dye bath. Adjust the desired M:L ratio. Then dyeing is continued. After 3–4 ends in color common salt is added in prescribed quantity to prevent the loosening of the naphthols from the substrate to the developing bath as well as to exhaust the base. The material is developed in developing bath up to 4–6 ends or as desired.

Finally washing, soaping is employed at boil.

3.3   Advantages of Naphthol dyeing

1. They are particularly strong in the orange, red and Bordeaux sectors; the range also includes dark blue and black.

2. Certain bright full depths are produced, not achievable by any other dye-class.

3. Excellent shade reproducibility is achievable.

4. Naphthol dyeing exhibits exceptionally good wet fastness on cellulosic fibers. All naphthol combinations withstand washing at boil. The light fastness is good to excellent in heavy dark shades.

5. The dyeing cost is low.

6. Most azoic combinations are fast to chlorine bleaching, but frequently inadequately fast to hydrogen peroxide.

7. With few exceptions, the fastness properties of azoic dyeing on mercerizing and perspiration are excellent.

8. The fastness to hot pressing varies from poor to excellent.

3.4   Disadvantages of Naphthol dyeing

1. The application procedure is complicated and time consuming.

2. The fastness of rubbing has always been a problem. High substantivity of the coupling component and thorough soaping is essential for good rubbing fastness.

3. There is limitation of hues available and shade matching is difficult, if not impossible. Each azoic color is developed by reaction of a diazo and a coupling component.

4. The final shade can be noticeable only after coupling. Hence any unevenness, which may have occurred during naphtholation, may be left unnoticed and is visible after coupling when shade correction is difficult.

5. They have poor light fastness in light shades.

6. The main weakness of Naphthol dyeing is their limited fastness to organic solvents used in dry cleaning and spotting.

7. Fastness to rubbing varies and is largely dependent on the dyeing process.

Azoic compositions of the Rapidogen color (mixture of naphthol and stabilized base) represent an alternative to indigo for the coloration of denim jeans. They can be applied to warp yarns on a simple two-dip sizing machine rather than a much more extensive indigo dyeing range.

The commercial products dealt with in azoic are those used to produce insoluble azo dye in situ, usually on a textile substrate .The basic principle of their application is the introduction of two small soluble components into the fiber and use of suitable conditions for coupling to occur, resulting in the production of insoluble colored molecule. One component is selected from the Azoic-coupling component and second from Azoic Diazo component.

Cotton, rayon, and other cellulosic fibers, as well as silk, can also be dyed with azoic or naphthol dyes. (Naphthol is sometimes also spelled as 'napthol' or 'naphtol'; the latter is the German spelling). Naphthol dyes are true cold-water dyes. The "cold" water used in fiber reactive dyes such as Procion MX dyes should, ideally, be between 35° to 41°C, although temperature as low as 21°C may be used. In contrast, naphthol dyes may be used in ice water. Both fiber reactive and naphthol dyes are suitable for use in batik, since they do not require heat that would melt the wax to set the dye.

While azo dyes are readymade dyes, azoic colors are formed inside the textile materials by reaction of two colorless or faintly colored compounds called naphthols, or CI Azoic Coupling Component and fast bases, or CI Azoic Diazo Components. At first, the naphthols are applied on textile materials by a process called naphtholation as they have some affinity for the cellulosic materials. The treated material is subsequently treated with a soluble diazotized form of the base, when intense color is formed. The later step is known as coupling or development. The solubilisation of the base is carried out with sodium nitrite and hydrochloric acid at low temperature and the process is diazotization.

Naphthol dyes are insoluble azo dyestuffs that are produced on the fiber by applying a Naphthol to the fiber and then combining it with a diazotized base or salt at a low temperature to produce an insoluble dye molecule within the fiber. Naphthol dyes are classified as fast dyes, usually slightly cheaper than vat *dyeings*; the methods of application are complex and the range of colors limited. The other name for Naphthol Dyes are Azoic dyes. The dyes containing insoluble azo group (–N=N–) are known as azoic dyes.

These dyes are not found in readymade form. Azoic dyes are produced by a reaction between two components.

The components are:

1. Coupling compound (Naphthol)
2. Di-azo compound or diazo base or diazo salt

Azoic combinations are still the only class of dye that can produce very deep orange, red, scarlet and Bordeaux shades of excellent light and washing fastness. The pigments produced have bright colors, and include navies and blacks, but there are no greens or bright blues. Crocking fastness varies with shades but washing fastness is equal to vat dyeings, generally with less light fastness than the Vats.

3.5   Azo coupling components (Naphthols)

The first practical azoic coupling component was β-naphthol. The azoic dyeing from β-naphthol had many limitations, such as the lack of substantivity of β-naphthol for cotton, the narrow range of shades obtainable from a single coupling component and the poor fastness to light and rubbing. In 1912, naphthol AS-arylamides of BON Acid was introduced as coupling components having definite cotton affinity and producing azoic dyeing of greatly improved brilliance and fastness. The group of azoic coupling components of this is known as the Naphthol AS series. These are generally made by condensing BON Acid with an aromatic amino compound with or without one or more of chloro, nitro, methyl, methoxy, etc., groups in different positions with respect to the amino group. Most of the naphthols are made from substituted aniline and naphthylamines.

The Naphthols are phenols, soluble in alkaline solution and substantive to cotton, particularly in the presence of salt. The anilides of BON acid (betaoxynaphthoic acid or BON acid) are soluble in dilute NaOH solution and form the corresponding naphtholate ion. These relatively small molecules are of only low to moderate substantivity for cotton, but they diffuse rapidly into the fibers. In general, the higher the substantivity, the better the rubbing fastness as less azo pigment forms on the fiber surfaces. The naphtholate ions are always coplanar and preferably have elongated molecular structures. They behave essentially as colorless, low molecular weight direct dyes. The substantivity increases with increase in the molecular size of the naphtholate ion, but the diffusion rate in the fibers and solubility in dilute aqueous alkali decrease. Addition of salt promotes better exhaustion of the bath, more being needed for Naphthols of lower substantivity.

Naphthols differ widely from one another in their affinity or substantivity for cotton fiber.

They can be classified as:

1. Low substantivity Naphthol
2. Medium substantivity Naphthol
3. Higher substantivity Naphthol
4. Still higher substantivity Naphthol

Substantivity of Naphthol can further be increased by adding electrolyte like common salt or Glauber's salt.

3.6   Bases

The second components used in the production of azoic are called azoic diazo components and are generally aromatic amines with various substituents selected from chloro, nitro, methyl, methoxy and other groups in different positions. These are made from the corresponding aromatic nitro compounds by reduction. These are marketed as fast color bases, either as the free bases or their hydrochlorides.

These are available as the free amine base or as amine salts such as the hydrochloride. Many of the amines used are simple substituted aniline derivatives with no ionic substituents. The so-called fast color bases require diazotization. This usually involves reaction of the primary aromatic amine in acidic solution or dispersion with sodium nitrite, at or below room temperature. Successful diazotization requires careful weighing of all the chemicals and regard for the supplier's recommendations. Diazotization of a primary aromatic amine is often difficult and solutions of diazonium ions are inherently unstable. They undergo decomposition even at low temperature and particularly on exposure to light. Storing prepared diazonium ion solutions is not usually possible.

The way naphthol dyes are used is fascinating. Two different types of chemicals are mixed in the fiber: the diazo salt and the naphthol; the specific combination determines the color obtained. An advantage of this sort of dye is that contrasting colors may be placed adjacent to each other on fabric without color bleeding from one to the other. As with vat dyes, the final color is provided by insoluble particles of dye that are stuck within the fiber; only the components that react together to form these compounds are themselves soluble in water.

Here is Richard Proctor and Jennifer Lew's chart showing how diazo salts can be mixed with naphthol bases to make different colors, from their book.

Table 3.1 Mixing different colors with Naphthol dyes.

	Base A Naphthol AS	Base B Naphthol AS.G	Base C Naphthol AS.GR	Base D Naphthol AS.LB	Base E Naphthol AS.BO
Salt "1" (Fast yellow GC)	red-orange	pink	red	bright blue	blue-violet
Salt "2" (Fast Scarlet R)	lemon	bright yellow	saffron	gold	ochre
Salt "3" (Fast Red B)	magenta	red-violet	purple	blue-green	green
Salt "4" (Fast Blue BB)	tan	chocolate	red-brown	purple	deep violet
Salt "5" (Fast Blue B)	bright red	deep red	maroon	blue	blue-black

3.7   General dyeing procedure of Naphthol dyes

The application of the naphthols consists of following steps:

1. Dissolution of the naphthol component.
2. Exhaustion of the naphthol solution onto the substrate or absorption of the naphtholate ion by the cotton.
3. Removal of excess naphthol from the material by squeezing, partial hydroextraction or brine washing.
4. Diazotization of the base component.
5. Development or treatment with the diazonium ion solution to bring about coupling.

6. Neutralization, soaping at the boil to remove superficial pigment, followed by rinsing and drying.

The process can be carried out in almost any type of dyeing machine determined by the form of the goods.

3.8   Dyeing methods

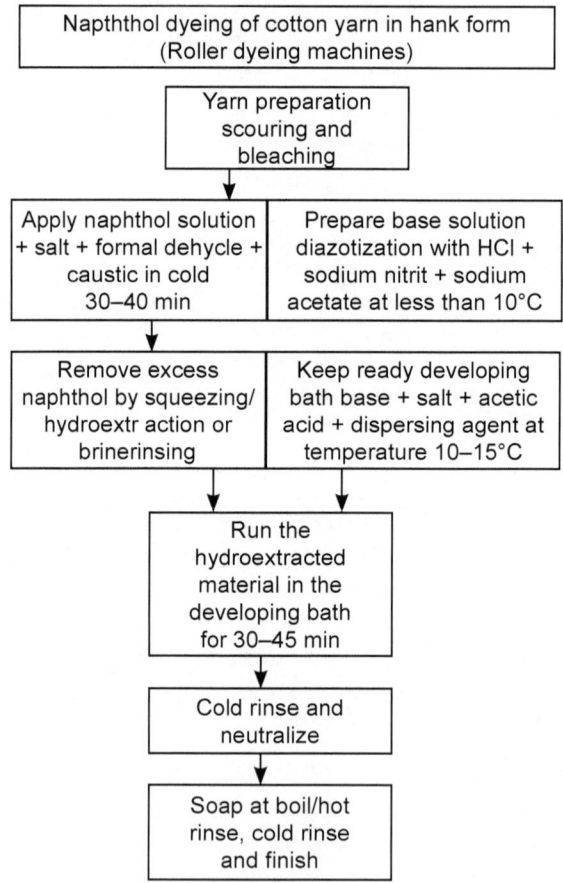

Figure 27

3.8.1   Precautions in Naphthol dyeing

1. The alkalinity of the naphthol bath shall not drop below the prescribed limit otherwise the naphthol may precipitate.

2. Formaldehyde shall not be used when working at more than 50°C or when the material is to be dried after naphthol application.

3. Material shall be protected from water spotting, steam, acid and chlorine fumes, and exposure to sunlight after naphthol application.

4. Use of excess salt in naphthol bath may result into precipitation of the bath.

5. The temperature is very important in base preparation stem, otherwise diazotization may not take place.

6. Sodium acetate must be added to the developing bath just before the use, otherwise base will become unstable due to fall in concentration of HCl.

7. Hydroextraction time must not be too long, which may result into light spots after development.

8. Material shall be rinsed without delay after developing, otherwise the mechanically held excess developing liquor will undergo some decomposition and cause deposition of dark colored spots, which will be difficult to remove.

9. It is important to use sufficient amount of alkali binding agents, otherwise it will result into precipitation of developing bath.

3.8.2   Stripping process in Naphthol dyeing

• Treat the dyed material with non-ionic detergent and 3–5 gpl caustic soda at boil for 15 min cool to 85°C

• Add 3–5% sod. hydrosulphite for 30–45 min at 85°C

• Rinse hot and cold

• Bleach with 1–2 gpl available chlorine for 20 min

• Antichlore and neutralise

• Soap at boil for 15–20 min

• Cold rinse

3.8.3   Roles of different chemicals in Naphthol Dyeing

T.R. oil – Wetting agents for naphthol pasting and dissolution and penetrating agent in fiber in naphthol application.

Caustic soda – For solubilising of naphthols and keeping proper alkalinity of naphthol bath.

Formaldehyde – Protective agent of naphthol impreganated material from effect of air.

Salt – Electrolyte for exhaustion of naphthol during naphtholation and to prevent the desorption of naphthol in the bath during brine rinsing and development phase.

HCl Acid – Dissolution of base and to produce nitrous acid in diazotization phase.

Sodium Nitrite – Producing nitrous acid in diazotization process.

Sodium Acetate – For neutralization of excess HCl in developing bath.

Acetic Acid – As an alkali-binding agent in developing bath.

Non-Ionic Dispersing Agent – To keep the azoic pigments in fine dispersion phase, which are formed by the coupling of free naphthol in developing bath. Also helps in better color fastness during soaping operation.

3.8.4   Fastness properties on cotton

Correctly prepared *dyeings with azoic* combinations on cotton have fastness properties often comparable, or only slightly inferior, to those produced using quinone vat dyes. They complement the vat dyes because of the wide range of orange, red and Bordeaux shades that they provide.

The fastness to washing of azoic combination dyeings on cotton is usually very good to excellent but only after careful elimination of particles of azo pigment loosely adhering to exposed fiber surfaces. Intermediate drying or rinsing of fabric containing the Naphtol, and the soaping of the final dyeing, are key processes ensuring optimum fastness. The same argument applies to rubbing fastness. Deep dyeing that have not been well soaped easily transfer color onto adjacent white fabric, even under conditions of gentle rubbing.

There are two other problems associated with the fastness properties of azoic combinations on cotton. In pale shades, the dyeings often have much reduced light fastness, particularly under humid conditions. Some sensitive azoic combinations also give dyeings of only fair resistance to chlorine and peroxide bleaching.

The Naphthol Dyes or Fast Base Dye is very widely used in the textile industry. Dyeing with Naphthol Dyes is more economical even if compared to Reactive Dyes. Maroon, Blue, and Yellow colors, obtained with Naphthol Dyes, are not possible with any other dyestuff. The fastness property of these Naphthol Dyes is excellent. Light Fastness is almost as good as 6–7, almost nearer to the Vat Dyes. The Naphthol Dyes are also used for African Print / Wax Print.

4

Sulphur dyes

4.1   Introduction

Sulphur dyes were first made in 1879. They are synthetic organic substantive dyes, produced by thionation or sulphurisation of organic intermediates containing nitro and amino groups and are customarily applied to cotton and viscose material from reduced bath containing sodium sulphide. They are like vat dyes, highly colored, water-insoluble compounds and have to be converted into water-soluble, substantive forms (leuco forms) before application to textile materials. The conversion is usually carried out by a treatment with dilute aqueous sodium sulphide solution.

Since these leuco compounds have affinity for cellulosic fibers and since they are sensitive to atmospheric oxidation (they are easily oxidized to the parent sulphur dyes), they have to be dyed from an aqueous sodium sulphide solution. Again, like vat dyes, the leuco sulphur dyes, absorbed by the fiber substance, have to be reconverted into the original water-insoluble dyes. This is generally carried out either by exposure to air ("airing") or by using a chemical oxidizing agent like sodium dichromate (one or two manufacturers produce sulphur dyes that have been made water-soluble).

Sulphur dyes have been in practical commercial use in most of the countries of the world for many decades. The appeal of sulphur dyes is due to their:

(a) Durability
(b) Flexibility of application
(c) High money value

These dyes are fast to washing, light and perspiration, but they have one weakness: excessive chlorine bleaching will strip the color.

The first sulphur dye appeared before direct dyes, but it was not until the evelopment of the sulphur blacks that this class of dyes made a major impact on cotton dyeing. Sulphur dyes now constitute the largest class of dyes in terms of quantity with an estimated worldwide production of more than 80,000 tons per annum, even though a different picture emerges if the calculations are made on the basis of price. They are the most economic dye class, yet for most of them the chemical structures are unknown.

They are normally distinguished in terms of the organic intermediates from which they are derived and the process of sulphurisation used in their manufacture. The characteristic feature of the dyes of this class is that they all contain sulphur linkages within the molecules. There is no black dye for cellulose, which can compare with the attractive shade and depth possible with C.I. Sulphur Black 1. It is one of the most widely used dyes in the world.

Sulphur dyes are important for black, navy, brown, maroon, olive and green colors in medium to heavy depths. The liquid brands are ideally suited for continuous dyeing in long terms.

Sulphur dyes are used widely on cellulosic fibers and their blends, especially with polyester, and also with nylon and acrylic. Cotton and polyester cotton drill and corduroy are dyed continuously or on the jig, whilst cotton–nylon and cotton–acrylic knitted and pile fabrics are winch and jet dyed. Since 1985, sulphur black finds increasing use on denim.

In the 1860s, chemists observed that colorants are formed when certain organic nitro compounds are reacted with alkaline sulphides. Nevertheless, the first commercial brown sulphur was made by Croissant and Bretonniere in 1873 by heating sawdust and bran with alkaline sulphides, the active component being lignin C.I. Sulphur Brown 1 is still manufactured from lignin sulphonates obtained as a by-product from sulphide bleaching of wood pulp for papermaking. Many of the sulphur dyes used today are made from well-defined chemical intermediates and a full range is available covering all hues except a true red.

Sulphur dyes are mainly used for dyeing cellulosic fibers for corduroy articles, work wears and men's outerwear (woven and knitted goods) and leisurewear. Polyester /cotton or polyester / viscose-blended fabrics are another important area of application of sulphur dyes. Silk, paper and leather are also dyed to limited extent. Sulphur dyes are much used for raincoat and other materials which are subsequently rubber coated and vulcanized or laminated with polyurethane foam. Of all the sulphur dyes, black, navy, brown olive and green dyes are the most important from the point of view of tonnage.

There is no Black on cellulose which can compare with the attractive shade and depth possible with C.I. Sulphur Black 1, which accounts for it being one of the most widely used dyes in the world. Sulphur dyes are generally faster to washing and light and are brighter in shade when applied to rayon than to cotton. They are important in producing a wide range of shades on a variety of cotton and rayon fabrics, especially on heavy, durable shades of apparel fabrics. They are dyed mainly on cheap cotton goods. As a class, sulphur dyes are fairly cheap among synthetic dyes. The shades are relatively dull and fastness to powerful oxidizing agents, especially sodium hypochlorite solutions is poor. This class is not complete in shade range. The true red in this class is not available.

It is relatively inexpensive to vats and superior in wash fastness when compared to reactives. The liquid brands are ideally suitable for continuous dyeing ranges.

4.2   Classification

As already discussed, sulphur dyes have no well-defined structure and hence they are best classified by their application methods. Sulphur dyes in commercial forms are available as powder, pre-reduced powders, grains, disperse powders, disperse pastes, liquids and water-soluble brands. There are now four main groups in the Colour Index classification:

(a) C.I. Sulphur dyes
(b) C.I. leuco Sulphur dyes
(c) .I. solubilised Sulphur dyes
(d) C.I. condense Sulphur dyes

(a) *C.I. Sulphur dyes* – These are water-insoluble dyes, containing sulphur both as an integral part of the chromophore and in attached polypeptide chains. They are normally applied on cotton in the alkaline-soluble reduced (leuco) form from a sodium sulphide solution and subsequently oxidized to the insoluble form in the fiber. These dyes are mainly used for exhaust dyeing method.

(b) *C.I. Leuco Sulphur dyes – These are* powders or liquids containing the conventional dye and reducing agent, normally in sufficient amount to make the dye suitable for direct application.

These dyes require no dissolving process. The reducing agent most frequently employed is mixture of sodium sulphide and sodium hydrosulphide and mixed with a slurry of raw dye paste which is then dried on a drum drier. The resultant product is really the leuco form of sulphur dye. They are best suited for single bath pad-steam process. Small additions of reducing agents should be made depending on the method employed for their application and the desired depth of color.

(c) *C.I. solubilised Sulphur dyes* – This class of sulphur dyes is obtained by reacting the conventional sulphur dye with sodium sulphite or bisulphate to give a thiosulphonic acid derivatives which are very soluble in water, but possess no affinity for the fibre until reduced in the presence of alkali. They are suited for exhaust and pad-dry-steam method of application. Cross-wound packages of yarn may also be dyed by two-stage method.

(d) *C.I. Condense Sulphur dyes* – They are S-alkyl or Sarylthiosulphates. Although they contain sulphur as their constitution and method of manufacture bear little resemblance to those of traditional sulphur dyes. Conventional methods of dyeing are unsuitable for this class of dyes.

4.3   Application

Sulphur dyes are versatile and can be applied by various methods. Some of the most popular methods are:

(a) *Jigger dyeing* – Prepare the bath with 0.25 gpl wetting agent, 2.5 gpl soda ash, 2.5 gpl anti-bronzing agent, 5.0 gpl sodium sulphide and 0.5 gpl sequestering agent and the fabric is wet out by two ends through the bath. Then reduced sulphur dye is added to the bath over two ends at 65°C (for black at 90°C).

Common salt is added for exhaustion of the dye over two ends and finally two more ends are run. The dyebath solution is dropped, washed with overflow water and oxidized the dye by chemical oxidation method. After dropping out the oxidation liquor, the material is washed with overflow water until clear and then soaped at boil at about 90°C, rinsed and unload.

(b) *Winch dyeing* – Fabrics like cut plushes, terry toweling and raised knitted goods are mainly dyed on winches or jet dyeing machines, in the rope form. The winch bath is set with 2.0 gpl anionic wetting agent and 2.0 gpl antioxidant and run for 10 min.

Raise the temperature to 60 deg c. Add the reduced sulphur dyes solution over 20 min. The bath is heated to boil within 20 min (for blue raise the temperature to 80°C). Common salt or Glauber salt is added in two installments over 10 min and dyeing continued for further 45 min. The goods are then rinsed thoroughly with overflowing water and oxidation is followed. Then cold wash followed by soaping at 70°C for 20 min. Then overflow rinse and unload.

(c) *Package dyeing* – The yarn is dyed in package form. Also cotton loose fibre dyeing is also carried in the same HTHP dyeing machines. The yarn soft wound on ss-dyespring or pp-perforated tubes is loaded in the yarn carrier. Similarly the carded or combed sliver is loaded in the fibre carrier. Dye dissolution is not required as nowadays sulphur liquid dyes are available in the market, which are already pre-reduced. The dyebath is set with 5 gpl Sodium sulphide, 2.5 gpl Soda ash, 2.5 gpl Anti-bronzing agent, 0.25 gpl wetting agent, 0.5 gpl sequestering agent. Run for 10 min. Add the diluted dye solution in two installments and run for 5 min each. Then increase temperature to 50°C in 15 min.

Add common salt or Glauber salt in two installments and run for 10 min. Raise temperature to 60°C (80°C for blacks). Run for 45–60 min. Then drain and overflow rinse. Oxidation is done at 60°C for 20 min using 2 gpl Hydrogen peroxide and 2 gpl Acetic acid. This is followed by rinsing, soaping and rinsing as usual.

(d) *Continuous dyeing* – In this system, the goods are padded with pre-reduced sulphur dye solution at 38°C except black shade. The padded goods are steamed at 102–104°C with saturated steam for 30–60 s, washed at 40–60°C, oxidized, soaped and rinsed.

(e) *Slasher dyeing* – Vast quantities of warp yarns for the dyeing of denim warp are dyed by this continuous technique. This system, which frequently employs modified sizing machines or 'slashers', is similar to that used for dyeing warp yarns in rope form. Instead of the warps being on balls in rope form, they are wound on a series of beams and the yarn in a 'sheet' runs continuously from the beams through a concentrated bath of reduced sulphur dye for 10–25 s at a temperature of 80–95°C.

The yarn is then passes over an airing frame and through a series of washing boxes before being dried. It is then sized, dried again and wound on the weaver›s beam.

(f) *Garment dyeing* – Sulphur dyes have found a wide acceptance among garment dyers because of their property of producing a good granular effect when after treated with biofading enzymes as well as due to their inherent poor fastness to chlorine which makes them a popular choice for garments which are to be 'acid-washed'.

4.4   After treatments

With the object of improving the fastness properties or stabilizing the shade, sulphur-dyed goods are often given after-treatments which are as follows.

(a) After dyeing and rinsing, the goods are treated with 1.5 gpl acetic acid to neutralize any alkali which may be remaining on the material after dyeing. Such type of treatment is necessary for C.I. Sulphur Red 6 to develop the red tone of the dyestuff.

(b) Especially for Blues to brighten the final shade, the goods are treated with 0.5 to 1 gpl of sodium perforate or hydrogen peroxide for 10–15 min at 40°C and finally rinsed.

(c) To improve washing fastness, the sulphur-dyed goods are treated with 1–5 gpl of potassium or sodium bichromate and acetic acid for 10–15 min at 60°C and finally rinsed.

(d) To improve the light fastness the sulphur-dyed goods are treated with certain metallic salts such as 1–5 gpl copper sulphate and acetic acid for 10–15 minutes at 60°C.

4.5   Common problems encountered and remedies

There are several problems of sulphur dyeing process. Few prominent defects are:

(a) *Bronziness of shades* – 'Bronziness' is caused due to premature oxidation. It generally occurs with the Navy and Blue shades. This may arise out of various reasons:

Excessive delay in lifting the material out of the dyebath. Insufficient amount of reducing agent in dyebath resulting in surface dyeing, where it gets quickly oxidized by air leading to bronziness. This problem can be minimized by treating the 'bronzy' dyed material in a bath containing 2–3 gpl Turkey Red Oil and 1–2 gpl Liquor Ammonia at 50–60°C for 10–15 min.

(b) *Tendering* – It means loss of strength of the dyed material, which occur during storage. It is particular in cases where large quantity of dyes are used such as Black, Navy shade etc. Generally there is some free sulphur associated with the sulphur black dye, in the manufacturing stage itself. This free sulphur is released during storage of the dyed fabric and combines with atmospheric conditions (excessive humidity) to form sulphuric acid which in turn tenders the fabric.

Tendering action can be minimized by giving an after-treatment with 2 gpl of Soda Ash or Sodium Acetate in the final bath and the fabric removed in the same condition without further rinsing ensuring the fabric slightly alkaline in the final stage.

Sulphur dyes are being used for dyeing of denim fabrics in blues and blacks. The characteristic rubbing fastness of sulphur dyes is advantageously utilized in technique like stonewash, sandwash, monkey wash, acid wash, brushing, etc., for getting better aesthetic effects. Research is going on about the feasibility of replacing the major polluting ingredients in the sulphur dyeing are the traditional reducing agent – sodium sulphide and the oxidizing agent sodium dichromate. Regarding these objectives, sulphur dyes manufacturer like SF Dyes Pvt Ltd., Sulfast Chemical Industries, Atul Ltd. and Clariant Ltd. are already working on these eco-friendly issues.

The exact chemical structure of sulphur dyes is not known, but these dyes contain sulfur as an integral of the chromophore as well as in the polysulphide side chains. These are produced by thionisation or sulphurisation of organic intermediates containing nitro and amino groups.

Sulphur dyes are compounds prepared by heating various nitrogenous organic materials with sulphur, sodium sulphide, sodium polysulphide or other sulphurizing agents. They are chemicals described as containing thiazole, thiazone and thianthrene rings with polysulphide linkages. The reaction is carried out in closed vessels in the presence or absence of solvent. Since little is known about the constitution of sulphur dyes, they are usually classified according to the chemistry of their starting materials and in accordance with their color.

4.6   Manufacturing of sulphur dyes

The two ways of manufacturing are:
1. Baking the organic material or intermediates with sulphur with or without sodium sulphide (Na_2S). Many yellow, orange and brown sulphur dyes are made by dry baking of intermediates such as 2,4-diaminotoluene with sulphur above 160°C.
2. Boiling the intermediates with sodium polysulphides, in aqueous or alcoholic solvents, under reflux conditions. The aqueous reflux method is used for the production of sulphur blacks, blue-green and violet dyes from indophenols or more stable leuco indophenols. Indamines on thionation produce reddish brown to Bordeaux dyes.

In both the cases, hydrogen sulphide is evolved during the reaction, and it is absorbed in caustic soda. The dye is precipitated by acidification or by oxidation or by both.

The properties of the products of sulphurisation are controlled by the intermediate selected, the condition of reaction such as time, temperature and the conditions of isolation of the dye from the reaction mixture. Sulphur dyes are among the cheapest of the synthetic dyes. Some knowledge of the structure of the very complex molecules of sulphur dyes is being accumulated gradually. For example, when sulphur is heated with p-toluidine, one of the reactions taking place is the formation of dehydrothiotoluidine. This repeatedly reacts with more and more toluidine to form a complex molecule. CI Sulphur Yellow 4 and CI Sulphur Orange 1 have a similar polythiazole structures with –S–S– bond with benzene rings.

Some of the sulphur dyes are assumed to be composed of several units of thiazone linked together by sulphur atoms. In the past, it was assumed that these links are invariably disulphide bonds, which are reduced to mercaptans (R.SH). However, now it is assumed that single sulphur bonds (–S–) between benzene rings also exist which provide substantivity to cellulose and survive reduction with sodium sulphide.

4.7   Properties of sulphur dyes

Chemically sulphur dyes are amorphous, colloidal materials of high molecular weight and variable composition.

These are water-insoluble dyes and have no affinity for the cellulosics as such, but solubilised when treated with a weak alkaline solution of sodium sulphide or any other reducing agent to form a leuco compound. These leuco compounds are water soluble and have affinity for the cellulosic materials such as cotton, viscose, jute and flex, etc. These dyes are absorbed by the cellulosic material in the leuco form from aqueous solution and when oxidized by suitable oxidizing agents, got converted into insoluble parent dye, which is fast to normal color fastness parameters.

Main properties of the Sulphur dyes are as follows:

1. Economical dyeing with excellent tinctorial value and good build up properties.
2. Good overall colorfastness properties such as wash fastness, light fastness, perspiration fastness, etc. Moderate fastness to crocking and poor fastness to chlorines bleaching agents such as bleaching powder and sodium hypochlorite.
3. The shades of sulphur dyes on cotton are limited in brightness and true reds are not available. Bright greenish yellows, bluish greens, blues and reddish blue of a good tinctorial value are available. Sulphur dyes are particularly rich in blacks of good tinctorial value and shades.
4. These dyes can be applied by exhaust, semi-continuous or continuous dyeing methods on garment, yarn, knits, fabric as well as loose stock, etc.
5. Available in powder, granules and liquid forms.
6. Sulphur black 1 is the major black dye used world vide for dyeing of cellulosics.
7. The conventional dyeing process is not environment friendly due to pollution problems of sodium sulphide as well as sodium /potassium dichromates.
8. When dyed by using non-polluting reducing and oxidizing agents, the process is environment friendly.
9. Sulphur dyes have good wet fastness, good fastness to bleeding in water, and to perspiration. An important weakness to sulphur dyeing is the lack of fastness to chlorine.
10. It has been suggested that the fastness of sulphur dyes to chlorine can be increased by treatment with a melamine–formaldehyde resin, which forms a chloramines derivative with chlorine.

4.8   Types of sulphur dyes

There are three classes of sulphur dyes, which are available commercially:

1. Conventional water-insoluble dyes which have no substantivity to cellulosics.
2. Solubilised sulfur dyes, which are water soluble and non-substantive to cellulosics.
3. Pre-reduced sulfur dyes, in the stabilized leuco compound form, which are substantive to cellulosics.

Sulfur dyes like vat dyes are applied to textiles (cellulose, Table 4.1) as a soluble anionic form and then oxidized into the insoluble form.

C.I. Sulphur Black 1 and C.I. Sulphur Blue 7 are amongst the most widely used sulfur dyes. Like other sulfur dyes, their structures are variable and largely unknown. They provide a range of blacks, browns and dull blues. They are however much cheaper than vat dyes to produce because their preparation by heating various organic compounds with sulfur is simple.

4.9   Application of sulphur dyes

4.9.1   Mechanism of the sulfur dyeing

The application of the sulfur dyes involves several steps, which are described as given below:

(1) *Dissolving the dyestuff* – The dye is taken in an SS vessel (size of the vessel should be selected as per the quantity and solubility of the dyes) and pasted well with a good alkali stable wetting agent and small quantity of soft water. A required quantity of soda ash may be added to neutralize any acid formed in the dyestuff during storage (if the acid is not neutralized, it will react with the sodium sulphide, resulting into formation of H2S gas, which will result into incomplete and poor reduction of the dyes). It is very important that the dye dissolution must be complete otherwise particles of the undissolved dyes may deposit on the surface of the substrate resulting into patchy dyeing and poor rubbing / washing fastness.

(2) *Reducing the dyes to form a Leuco Compound* – Chiefly sodium sulphide is used as a reducing agent for the sulfur dyeing. The quantity of the reducing agent depends upon the shade depth and M:L of the bath. For complete reduction the required quantity of the sodium sulphide is dissolved in a separate container and solution is allowed to settle for 10–15 min before decanting the clear solution into the dye-dissolving vessel. Further boiling water is to be added to make up the required volume, then heated to boil for

10–15 minutes either by live steam or indirect heating, for complete reduction of the dyestuff.

3. *Dyeing with the reduced dyes* – It is advantageous that the goods are scoured well before dyeing to have a satisfactory absorbency for better penetration. The dye bath is kept ready with small quantity of the alkali stable and a compatible wetting agent, a dye bath stabilizer, sodium sulphide and caustic soda or soda ash to maintain the alkalinity of the dye bath. The dye solution is then added through a filter cloth slowly over 15–25 minutes and then run for another 15 minutes at 40–50°C, then temperature is raised to 60°C and electrolyte is added in at least 3 portions. The quantity of salt added depends upon the type of shade, depth and dyestuffs; however a maximum quantity does not exceed more than 15 gpl. The temperature is then raised to above 80°C or even boil depending upon the dyes and kept for sufficient time to get the desired shade.

After getting the correct shade the bath is either dropped by draining the contents or by collecting it in the storage tanks for reuse after replenishing with fresh dyestuffs.

4. *Washing off the unexhausted dyestuff* – With an objective of achieving the highest possible color fastness results such as washing, rubbing, light and perspiration, the material is washed and rinsed several time with fresh water to remove maximum possible loose residual dye as well as sodium sulphide from the material. At the end of the washing process the water should be clear, with no further leaching out color. After washing the material is given a hot wash at 70°C.

5. *Oxidation back to the parent dye* – The oxidation is done to reconvert the leuco compound back to insoluble parent dye. There are number of methods available for oxidizing the leuco compound which are used either independent or in combination, such as

    (a) Oxidation by exposing the dyed material to atmospheric oxygen.

    (b) Oxidation by the dissolved oxygen in the fresh water.

    (c) Chemical oxidation, by employing different oxidizing chemicals, such as

        (i)   Acetic acid

        (ii) Sodium perborate in cold at neutral pH

        (iii) Hydrogen peroxide and acetic acid

        (iv) Potassium or sodium bicarbonates and acetic acid

6. *After treatment* – After oxidation and hot wash, the material is neutralized with soda ash to adjust the pH and then soaping treatment is done with a neutral soap and soda ash at boil. Followed by a hot wash at 85°C.

7. *Dye fixing treatment* – Optifix F (clariant) is a cationic dyefixing agent, which is applied in alkaline conditions (at a pH of 10–11), and is a suitable dyefixer for sulphur-dyed material to improve the color fastness.

8. *Softening* – A suitable (compatible) softener can be applied to the dyed material as per the intended end use and dyestuff applied.

9. *Final treatment* – To avoid the tendering of the dyed material, final wash is given to maintain a slight alkaline pH by a weak base or acid neutralizing agent at the end without further washing. Following treatments are recommended:

(a) Soda ash wash 2–3 gpl

(b) Sodium Acetate 2–3 gpl

(c) Tetrasodium pyrophosphate 5.0 gpl

(d) Lime and tannic acid treatment

10. *Use of standing bath* – Since a large quantity of the dye is always present in the unexhausted form in the spent liquor, this remaining dye can be reused after replenishing with fresh dye. This system is particularly suitable when producing repeated lots of the same shade with a single dye, such as black.

The dye liquor at the end of dyeing cycle is collected in the tanks, to replenish the bath a separately made dye solution is added and calculated quantities of sodium sulphide, soda ash as well as salt are added. The final volume is made up to the required level and reused. Usually a 50 –70% dye is replenished in case of blacks.

The spent bath use is not recommended in case of mixture shades, due to difference in the exhaustion and fixation of individual dyes.

4.9.2   Common problems and corrective action

1. *Poor wash and rubbing fastness* – Poor washing and rubbing fastness is generally caused by improper color dissolution, color precipitation, poor solubility of the dyes, poor and insufficient washing after dyeing of unexhausted dyes and poor or insufficient soaping treatment. To get overall good fastness properties:

(a) The dye dissolution must be complete and it should be filtered before adding to the dye bath, because insoluble dye particles, if present, will stick at the outer surface of the substrate causing unleveled dyeing and poor wash and rub fastness.

(b) The color should be dissolved in sufficient quantity of water, by keeping in mind the maximum solubility of the dye.

(c) The water and the salt should be free from calcium and magnesium, which, if present will make insoluble inert salts, which precipitates especially in the closed dyeing machines, in the form of sludge.

(d) The washing after dyeing and soaping treatment must be efficient to clear all the unused dye as well as chemicals, before going to the next operation such as oxidation and neutralization respectively.

2. *Bronziness* – There are various reasons for bronziness in the sulphur dyed material such as, in sufficient quantity of sodium sulphide or reducing agent, resulting into quick oxidation of surface dyeing. The presence of excess dyestuff on the material caused by high concentration of dye or electrolyte, delay between dropping of bath and washing, oxidation step. Following are the corrective actions for correcting and avoiding the bronziness problem:

(a) Proper dissolution of the dyestuff.

(b) Thorough washing and treatment with reducing agent before oxidation.

(c) Use of surfactants, sequestering agents, dispersing agents, dye bath stabilizers, and anti-oxidants in reducing bath.

(d) Using sufficient and calculated quantity of reducing agents.

(e) Using appropriate quantity of electrolyte e.g. less than 15 gpl.

(f) After treatment with 2–3 gpl TR oil+ 1–2 cc/ltr of ammonia in luke warm bath, to overcome the problem.

(g) Treatment with soap solution at boiling temperature.

(h) Using a blank bath of sodium sulphide.

3. *Tendering* – Tendering means the loss of strength or degradation of cellulosic materials upon storage. The tendering is caused by the acid formation from the free sulphur present in the dyed material by the action of moisture and air. The acid produced reacts with cellulose and degrade it, resulting in loss of strength. The tendering can be minimized by giving after treatments with acid neutralizing agents or by weak alkaline washing at the end of dyeing process.

4. *Poor color value* – Poor color value is caused by insufficient amount of reducing agent, presence of calcium salts in water and salt, over reduction of dyestuff, over oxidation, etc.

5. *Correction of faulty dyeing* – If the dyeing results are unlevel, then these can be corrected by:

(a) Leveling the dyed material by running in a blank bath containing excess sodium sulphide, dispersing, sequestering agent, wetting agent at a temperature of 80–90°C, this treatment will partially strip the color, which can be adjusted in a fresh bath. Or alternatively the partial stripping can be done by using caustic soda 5 gpl and hydros 5 gpl at a higher temperature than the dyeing temperature.

(b) For poorly leveled material, the material is treated with sodium or calcium hypochlorite, in which it is treated with 2–3 gpl available chlorine at room temperature, followed by thorough wash and neutralization and antichlore treatment.

4.9.3   Water quality for sulphur dyeing

The use of soft water with less than 50 ppm hardness is preferred which should be free from calcium salts, but in case only hard water is available, a sequestering agent based on sodium hexametaphosphate or EDTA should be used. These chemicals avoid the formation of insoluble metal–dye complexes which cause poor rubbing fastness and uneven dyeing.

4.9.4   Other recommended chemicals in dyeing

(1) *Wetting agents* – Normally 1–2 gpl wetting agent is used for good penetration, in the dyeing bath. Wetting agents used must be compatible with the dyestuff, particularly in combination shades. The wetting agents must be low foaming and alkali stable at high temperature. Unsuitable wetting agents adversely affect the dye bath, inhibiting the dye uptake or precipitating the leuco compound of the dye. Normally 1–2 gpl of wetting agents are used in the dyeing bath for good penetration.

2. *Dispersing or dye bath conditioner* – These are used to impart the leveling effect as well as to keep the dye in dispersed form, to avoid the dye aggregation and precipitation. Generally naphthalene sulphonic acid–formaldehyde condensate, lignin sulphonates and sulphonated oils are used in sulphur dyeing.

4.10   Dyeing of liquid sulphur black

Black is one of the highest volume shade dyed on cotton & synthetic textile material having all time great demand especially for casual wear (denims & garments). Amongst all the classes of dyestuffs, Sulphur black is an important class of dye for the coloration of cellulosics, being into existence for nearly a hundred years. Sulphur dyeing is used for dyeing of cotton and in market it is available in powder and liquid form also. In exhaust dyeing maximum powder form is used in dyeing of cotton material. In sulphur black dye, main advantages are: it can cover immature cotton available in fabric portion and get lot of fashion shade. Even in Fabric Dyeing general fashion is there if shade wise any major problem is there convert it sulphur black shade. The good fastness properties, cost effectiveness & ease of applicability under different processing conditions exhaust, semicontinuous and continuous make it one of the most popular dyestuffs. Further, a wide choice of selection of various forms conventional, leuco and solubilised form is the major factor contributing to the continuous existence & ever-increasing demand for this class of dyestuff.

Basically Sulphur dyes belong to reduction oxidation-dyeing system. Though Sulphur Black is a leading member, demand for other colors like Greens, Navy Blue, Browns, Bordeaux, Khakhi & Olives is also increasing. The environmental pollution problems associated with conventional dyeing systems involving use of Sodium Sulphide as a reducing agent is addressed by use of a non-polluting system consisting Glucose + Dithionate resulting in ecological advantages. Sulphur Blacks are available in grains, ready-to-use liquid form as well as solubilised form.

The liquid sulphur dyes are distinctly different from conventional sulphur dyes. These products are alkaline solutions of reduced sulphur dyes. The dye molecules are solubilized by adding suitable reducers in an alkaline medium. All products are practically sulfide-free (less than 0.3%). This very low amount of sulfide prevents the danger of undesirable release of hydrogen sulfide during acidification. No additional safety precautions are necessary while using liquid sulphur dyes. They give high coverage and good light fastness whilst wastewaters remain practically clear. Liquid sulphur black dye is considered better than solid sulphur dyes. Compared with solid dyes, these are more eco-friendly and have excellent osmosis, excellent equalizing, bright paint and convenient usage.

The features are:

1. Vivid dyeing color
2. Even dyeing
3. Simple craftwork and use
4. Use & stable quality
5. To get the best reducing state, stable and excellent color development effect, the course of production is controlled by strict technical rules; the reducer and dosage are filter and calculated carefully
6. Comparing with vat dye and reactive dye, this product has the features of low cost, less dyeing time, and convenient process

7. They are environment and mankind friendly, no harmful and carcinogenic substance inside.

8. Comparing with the sulphur dye in powder, the liquid sulphur black dye has more excellent osmosis, brightly painted, excellent equalizing, simplicity of operator, convenient to use.

9. This product has the features of low cost, less dyeing time, convenient process.

The specifications are:

1. Product: Liquid Sulphur Black
2. C.I. No.: Sulphur Black 1(53185)
3. Appearance: Black Viscid Liquid
4. Strength: 100%
5. pH Value: 12.0–13.5
6. Viscosity (mpa.S): < 300
7. Density (g/cm^3): < 1.30–1.50

Properties of liquid black sulphur dyes

The benefits of process are as follows:

1. *Convenience of use:* No dissolution of dyes required as ecosols are pre-reduced stabilized sulphur liquid dyes containing carefully controlled amount of reducing agent in order to obtain best stability in storage and color yield in use.

2. *Redyeing cost saving:* Redyeing will be practically nil/ negligible since improper reduction of dye is eliminated.

3. High productivity due to shorter dyeing time.

4. Good all round fastness.

5. Compatibility of dyes for combination shade in all proportion with each other.

6. Free from banned Amines and in conformity with Euro norms.

7. Eco-friendly process: Since there is very low salt addition, there would be lesser load on effluent discharge and treatment.

8. Superior reproducibility & consistency: Black shade produced will have better reproducibility and consistency.

9. Process time saving: There may be some reduction in process time.

10. Easier washing off when compared to reactive.

11. Good coverage of dead cotton.

The pre-reduced sulphur dyes are available as stabilized liquids, which have substantivity for the cellulosic, materials.

4.11   Main properties of liquid sulphur dyes

(a) No dissolution required, therefore cleaner environment and working conditions

(b) No use of sodium sulphide, therefore lesser smell

(c) Low salt additions are required

(d) Lesser pollution loads

(e) Easier washing off of the reducing agents, therefore easy oxidation

(f) Less staining and contaminations of the dyeing machines

(g) Good storage stability and water solubility

These dyes are recommended for exhaust dyeing of cotton in loose fiber, yarn, fabric and continuous dyeing, such as rope dyeing.

The use of non-polluting chemicals and by reusing the spent liquor dye bath, the dyeing becomes less polluting and environmental friendly. There are two major pollutants generated in classical sulfur dyeing procedure:

(a) Sodium sulfide in the reducing step

(b) Potassium/ sodium dichromate in the oxidation step

Both these chemicals are potentially hazardous for the environment, but can be replaced by environment friendly, less polluting chemicals such as, for reducing baths,

(1) Sodium sulfhydrate and alkali (soda or caustic)

(2) Sodium hydrosulphite and caustic soda

(3) Sodium hydrosulphite in glucose/caustic

(4) Glucose and caustic soda

(5) Alkaline sodium formaldehyde sulphoxylate

For oxidation baths

(1) Hydrogen peroxide and liquid ammonia

(2) Sodium perborate

(3) Sodium bromate and acetic acid

(4) Alkaline solution of sodium chlorite at pH 10

(5) Air oxidation, wherever possible

4.12   Use of spent dye bath in dyeing

A major consumption of the sulfur dyes is in the dyeing of black shades, and a large amount of dye is used to produce a good black. Due to high concentration of dye in the dyeing bath, the dye is not transferred to the substrate and a large amount of dye always remains unexhausted at the end of dyeing. Which if drained creates problem at water treatment plants and increases the cost of treatment. The unexhausted dye in these cases can be reused, after replenishing with fresh dye, when repeated lots of a particular shades has to be produced (say black).

The dye liquor at the end of dyeing cycle is collected in the tanks made for this purpose, the volume is made up for the lost liquor in dyeing, the dye which is to be replenished is separately added to it. Similarly the quantities of the electrolytes, and reducing agents are calculated and replenished. This bath then can be used as a fresh dye liquor.

The spent dye bath reuse is recommended for the self shades and blacks only, and not in combination shades because where a mixture of dyes is used the exhaustion properties of the dyes is different and it is not possible to replenish the bath, for producing the exact shade.

4.13   Identification of sulphur-dyed material

Take small specimen (about 3 cm × 3 cm) of dyed fabric in 25 ml distilled water, add 1 to 2 ml of 5% sodium carbonate (Soda Ash) solution, 500 mg of solid sodium sulphide and boil for 2 min. Remove the specimen. Add 25 mg of sodium chloride and a few pieces of white bleached cotton. Boil for 2 minutes; expose it to atmosphere by placing it on a filter paper along with test sample. If bleached cotton is dyed (except for strength) to the same color as that of test sample then it proves that it is sulphur dye.

4.14   Identification of sulphur dye powder

Take about 1 g of dye powder in a test tube and add small quantities of stannous chloride solution and hydrochloric acid. Close the test tube with a filter paper moistened with lead acetate. Heat the test tube till the solution boils. Appearance of brown stains on the filter paper indicates that the powder is sulphur dye.

4.15   Stripping of Sulphur dye from the dyed fabric

This can be done by treating the dyed goods in a blank bath containing 6 g per liter sodium sulphide at high temperature. A blank dye bath at 90–95°C containing 4.5–9 g per liter sodium hydrosulphite (hydros) and 2–3 ml per liter sequestering agent will strip about 10–20% color while addition of a stripping agent such as polyvinyl pyrrolidine will remove a further 10–20%.

DA-BS800 can strip sulfur dyes also. The dyeing matters can be stripped in bath of sodium hydroxide and sodium hydrosulfite, but can't in boiling bath of sodium sulfide. The stripping method or application rule of sulfur dyes is same to vat dyes.

5
Vat dyes

5.1   Introduction

The use of vat dyes as coloring matter can be traced to ancient times when people used natural animal and plant derivatives to color garments and other textiles. Some of the earliest references to such coloring substances go back thousands of years when Tyrian Purple was used. The dye was of animal origin, being an extract from shellfish. Indigo was the most important coloring material of the age. It has been known in India since the earliest periods for which historical records last.

These dyes are applied by chemical reduction. Before chemical reducing agents were available, the naturally occurring vat dyes were reduced by fermentation of organic matter in a wooden vessel referred to as a vat – and this is the origin of the name vat dyes.

One of the disadvantages of natural indigo was that the proportion of indigo, indirubin and analogous products were never constant, resulting in poor reproducibility of the dyeing. The manufacture of the substance from coal-tar derivatives constituted one of the classical triumphs of industrial organic chemistry. Indigo was first synthesized by Adolf von Bayer in 1880 and was considered as a successful replacement of natural product. BASF started industrial production of synthetic indigo in 1897. Karl Heumann prepared synthetic indigo by the phenylglycine route in 1890 and Hoechst started production of indigo in 1898 by adopting sodamide method of cyclisation. The successful large-scale production of indigo started from about 1900.

Subsequently, the synthesis of vat dyes has been an extremely active field of research and many new members of various complex chemical structures were incorporated into this dye-class. The industrial use of vat dyes started at the beginning of 20th century. By 1966, vat dyes accounted for about 26% of all the dyes marketed in the USA and 40–50% of all dyeings on cellulosic fibers. An estimate made in 1979 showed vat dyes to have a 15% share with indigo contributing a further 2%. Vat dye is so-called because indigo was dyed on textiles in wooden vat in ancient days. The old-fashioned method of fermenting and steeping indigo in a vat gave vat dyes their name. They are prepared today on the same principle, but they are chemically purer and the process is shorter. The chemical nature of vat dyes and their method of application (solubilisation by reduction, rapid penetration into the fiber followed by oxidation into insoluble color) guarantee an exceptional standard of fastness. However, the number of dyes introduced in the range has not increased much after 1968. A survey in Europe showed that out of 207 dyes manufactured in 1984, only 19 dyes were developed during the period 1968–84.

Vat dyes are the fastest dyes for cotton, linen, viscose rayon, blends and union (warp and weft of different fibers) fabrics containing above cellulosic fibers. They also may be applied to wool, nylon, polyesters, acrylics and modacrylics with the use of a mordant. Vat dyes are not only resistant to light and to acids and alkalies, but are also equally resistant to the strong oxidizing bleaches used in commercial laundries. In this respect, vat dyes excel sulphur dyes, which are not fast to chlorine washing. Vat dyes are expensive because of the initial cost as well as the method of application. This most important dye-class for cellulosic materials provides excellent all-round fastness properties, which cannot be attended, by any other dye class. This applies not only to their fastness in use, e.g. fastness to washing and light, but also to those factors important during fabric processing after yarn dyeing, e.g. fastness to mercerizing, soda boiling, chlorine and hypochlorite bleaching. Vat dyes are, therefore, ideal for colored woven goods such as handkerchiefs, shirtings and toweling. They are used for all outlets where the highest level of fastness is required such as:

(a) Work wear and uniforms for armed forces, police, fire, nursing services, etc., which are subjected to severe laundry washing and occasional bleaching with hypochlorite and long exposure to sunlight.
(b) Clothing and household articles such as shirtings, sportswear, toweling, bed sheets, tablecloths.
(c) Furnishing requiring high fastness to light.
(d) Outdoor fabrics requiring high weather fastness such as parasols, tenting, tarpaulins.
(e) Yarns like sewing threads and for colored threads for weaving which are subsequently piece – bleached with hypochlorite.
(f) Yarns used for border of dhoti, towel, etc.
(g) Shirting fabrics with white stripe go to mercerizing in fabric stage after weaving with dyed and grey yarn.

Vat dyes are water-insoluble pigments and are available in three forms:
1. Powder form
2. Paste form
3. Liquid form

Powder form is more finely dispersed form and is easily vatted. This type of dyes contain slight large size particle so recommended for exhaust dyeing. Their dye concentration is very high. Microdisperse (MF) and ultra disperse (UD) dyes have extremely high uniformity and produce good and stable dispersion. Its particle size is very small and fine, so recommended for padding as well as exhaust process.

Paste brand of dye is mainly used for printing. Dyes in liquid form have proved to be very advantageous; they facilitate metering – particularly when used in large amounts. The vat dye is found amongst the oldest natural coloring matters used for textiles. Indigo has been known in India since the earliest periods of which historical records exist. The naturally occurring vat dyes in plants, such as woad and indigo are known to be used 4000 years ago. Since antiquity, these, along with Tyrian purple extracted from shell fish, were the vat dyes applied by natural fermentation to cellulosic fibers. Vat dyes primarily are used to produce exceptionally fast colorations on any cellulosic fiber, chiefly cotton. However, under proper conditions, they will dye to some extent almost any fiber (like silk, wool, acrylic, nylon and polyester fibers) with varying degrees of fastness. Yield and fastness vary substantially, thus limiting considerably the practicability of commercial application. In 1897, von Baeyer first synthesized indigo, which was the start of rapid progress in the discovery of many vat dyes, such as Indanthrene which was synthesized four years later.

According to the writings of Julius Caesar, the ancient Britons used woad to stain their bodies and faces and Tyrian purple was exported from Tyre to the Mediterranean countries nearly 4000 years ago. Tyrian purple is extracted from a shell fish and is therefore of animal origin but woad and indigo exist in plants, combined with glucose in the form of glycosides the vat dyes are all insoluble in water and cannot be used for dyeing without modification when treated with reducing agents. They are converted into leuco compounds all of which are soluble in water in the presence of alkalis. These leuco compounds are substantive towards cellulose and reoxidize to the insoluble colored pigment within the fiber when exposed to air. The leuco compounds are often colorless or of quite a different color from the product of oxidation.

Despite their relatively high cost and difficult methods of applications, anthraquinone vat dyes are one of the most important groups of synthetic dyes because of their all-round superior fastness. One of the earliest of these dyes was blue vat indanthrone (CI Vat Blue 4). It was prepared in 1901 by the cyclization of Beta Aminoanthraquinone and chloroacetic acid with caustic alkali, in an attempt to obtain an anthraquinone analogue of indigo. Subsequently, a wide range of shades have become available in this series. The anthraquinoids are rich in blues, greens, browns, khakis, and blacks, but a serious defect of many of the yellow and orange dyes is their property of accelerating the degradative action of light and bleaching agents on cellulose. The vats undergo characteristic coloration in sulphuric acid and nitric acid, as well as in concentrated sulphuric acid upon the addition of small amounts of potassium persulphate or divanadyl trisulphate. The dyed fiber, treated successively with acidified permanganate and hydrogen peroxide, undergoes color changes that are useful as supplementary tests, particularly for certain groups of dyes such as the halogenated indanthrone and benzanthrone derivatives.

An estimate made in 1979 showed vat dyes to have a 15% share in the dyeing of cellulosic fibers, with indigo contributing a further 2%. The chemical nature of vat dyes and their method of application (solubilisation by reduction, rapid penetration into the fiber followed by oxidization into insoluble pigment) guarantee an exceptional standard of fastness. For certain sectors, it would be inconceivable not to use vat-dyed qualities. However, the number of dyes introduced in the range has not increased much after 1968. A survey in Europe showed that out of 207 dyes manufactured in 1984, only 19 dyes were developed during the period 1968–84.

This most important dye-class for cellulosic materials provides excellent all-round fastness properties, which cannot be attended, by any other dye-class. This applies not only to their fastness in use, e.g. fastness to washing and light, but also to those factors important during fabric processing after yarn dyeing, e.g. fastness to mercerizing, soda boiling, chlorine and hypochlorite bleaching. Vat dyes are therefore ideal for colored woven goods such as handkerchiefs, shirtings and toweling.

If the dye is readily reducible, poor fastness ratings may be obtained while washing under alkaline conditions at boil. This applies to the flavanthrones (C.I. Vat Yellow 1), the pyranthrones (C.I. Vat Orange 9) and to some extent, the dibenzathrones (C.I. Vat Blue 20). The problem can be prevented by adding a mild oxidizing agent, e.g. sodium m-nitrobenzenesulphonate. The indanthrones, especially C.I. Vat Blue 4 are sensitive to hypochlorites.

The name Vat was derived from the large wooden vessel from which vat dyes were first applied. Vat dyes provide textile material with the best color fastness of all the dyes in common use. Vat dyes are an ancient class of dye, based on the original natural dye, indigo, which is now produced synthetically, and its close chemical relative, historic tyrain purple. Cotton and wool, as well as other fibers, can be dyes with vat dyes. Vat dyeing means dyeing in a bucket or vat. It can be done whenever a solid even shade, the same color over the entire garments, is wanted, using almost any dye, including fiber reactive dye, direct dye, acid dye, etc. The opposite of vat dyeing is direct dye application, such as, for example tie-dyeing "vat dyes" are a special class of dyes that work with a special chemistry. Dyeing with vat dyes on cotton is a very old process.

Due its high fastness property it is preferred. Till the 70s vat dyes had a wide application. But in 80s reactive dyes have taken over from vat dyes due to its ease of application and cost effectiveness, thanks to technological development of reactive dyes having better washing fastness properties. But still then whenever chlorine fastness is considered, only vat dyes can pass the stringent tests. BASF was the first company to introduce vat dyes in India. It was the Indigo dyes. Used for denim dyeing.

Later on IDI (Indian dyestuff industries, Mumbai) came out with full range of vat dyes, then ATUL /ATIC, valsad, Gujarat started manufacturing vat dyes. INDOCHEM, Mumbai, also came out with vat dyes. Nowadays vat dyes are coming from China. Few Indian companies have entered the vat dyes market like Meghmani dyes, Ahmedabad; Colortex dyes, Surat; and Rathi dye chem, Pune. As compared to older days the vat dyes rates have comedown but still it is very costly as compared to reactive dyes. In those days vat dyes were used in dyeing 100% cotton fabrics on jigger dyeing machines. Old technicians were masters in vat dyeing. Then vat dyes were used in polyester/ viscose, polyester /polynosic or polyester/cotton blended fabric dyeing. Then vat dyes started using in package yarn dyeing.

It was also used in continuous dyeing range. Lastly in loose fiber dyeing vat dyes are also used.

Application was difficult but gradually the technicians had adopted the process tactfully. Still then the new generation technicians are not comfortable in vat dyeing as compared to reactive dyeing. Especially the cotton yarn dyeing the comfortable level is not there. In the 80s flax, ramie fiber was dyed using vat dyes in Jayashree Textiles, Rishra, Hooghly, West Bengal. Then worsted blends were made such as polyester/wool/flax like 70/20/10 or polyester/wool/ramie with the same blend.

Similarly in Gujarat Spinners Limited, Amletha, Bharuch, Gujarat. Coarse denier viscose was dyed using vat dyes like 6 denier, 8 denier, 10 denier etc. and blends were polyester/viscose/flax like 65/30/5. These were known as fancy yarn. In RSWM process house at Mordi, Banswara, Rajasthan, they were doing job work for Digjam in which they used vat dyes for dyeing the viscose part in polyester/viscose suitings. Similarly in Raymonds, Thane, for dyeing of viscose part vat dyes are used.

It is costly but they charge extra cost for va-dyed product. Morarji Brembana Mills, Nagpur, they make 100 % cotton shirtings for exports. The design is stripe/checks with white background.

The yarn is dyed in package form using vat dyes in HTHP dyeing m/cs and the dyed yarn is woven with grey 100% cotton yarn, then the fabric is mercerized and finished. The vat dyes are fast to mercerization. No color change. Vat dyes are most suitable for the dyeing and printing of cotton, linen and viscose fabrics. In some shades of nagda dope dyed viscose properties wise comparable with vat dyes. Because they possess high affinity for both natural and regenerated cellulosic fibers to which no other classes of dyestuffs can be compared.

If you are buying a vat dyed shirt /pant, the shade is brilliant with all round good fastness properties, particularly the light fastness and soaping fastness are more outstanding. Light fastness results are as per AATCC 16 E (source Xenon Arc) at various depths with 20 hours exposure. The rating was 4–5 assessed against 1–8 Standard scale. Vat dyes are derived from Anthraquinone. They are marketed in various physical forms/ grades like acraconc, powder fine, microperle and supra disperse. These differ in tinctorial strength and particle size. They can be applied to cellulosic substrate – fiber, yarn and fabric.

They are supplied in water-insoluble form. When applying them to cotton/viscose/linen, it is necessary at some stage of processing to convert them into their soluble leuco state by means of sodium hydrosulphite or hydro (reducing agent) in the presence of caustic soda. The leuco form has high affinity for cellulosic fiber. Once exhausted, the leuco vat is converted back to original form inside the fiber by oxidation. Otherwise at the end of dyeing process, the insoluble form is regenerated by an oxidation process. Powder and acra form are recommended for batch-wise leuco dyeing of cellulosic material. But supra disperse grade forms stable dispersions having uniform particle size capable of fast rate of reduction to fulfill the varying needs of dye applications like jet dyeing m/c s, HTHP dyeing m/c s and winch dyeing m/c s. They are also suitable for leuco methods and also for pre-pigmentation methods in exhaust dyeing.

Vat dyes are applied by three different methods designated as Method no. 1, Method no. 2 and Method no. 1 special. The methods, however, do not differ fundamentally and many dyes can be applied by more than one method.

Following is the classification of vat dyes (supra disperse) as per methods:

1. *Method 1 dyes* – Yellow 5G, N. Blue RA, Blue RCL, Blue BC, Jade Green XBN, Olive Green B, Grey 3B.

2. *Method 2 dyes* – Brown R, Brown BR, Olive R, Yellow 3R, Golden Orange 3G, Red 3B.

3. *Method 1 (special) dyes* – Direct Black AC.

Accordingly, chemicals to be used are mentioned as below:

1. *Method 1* – Caustic soda flakes – 5–8 gpl depending on light/med/ dark

    Hydro – 2.5–5 gpl depending on L/M/D shades

2. *Method 2* – Caustic soda flakes – 2–5 gpl depending on L/M/D shades

    Hydro – 2.5–5 gpl depending on L/M/D shades

    Common salt – 5–15 gpl depending on L/M/D shades

3. *Method 1 spl* – Caustic soda flakes – 12–15 gpl depending on med/ dark shades

    Hydro – 7–8.5 gpl – depending on med/dark shades

The above concentrations of chemicals are recommended as a guide for dyeing at a liquor ratio of 10:1. Adjustments have to be made at the plant level to adapt to particular conditions of liquor flow, aeration in addition tanks and so on. Please note when working at short liquor ratios, the concentrations of chemicals should be increased slightly, say by 10–15%. Similarly, lower concentrations may be used at longer liquor ratios. For the production of compound shades, it is advisable, wherever possible, to use components that are normally dyed by the same method at the same temperature as a general rule. The method to be adopted for dyeing combination shades will be governed by the proportion of the dyestuff of different methods of dyeing present in the mixtures and by their dyeing properties.

Vat dyes pasting has to be done using the dispersing agent Setamol WS (BASF). To obtain level dyeing in vat dyes, scouring should be very proper. Scouring is done at boil for 30 min using 2 gpl Lissapol D paste (ICI detergent) and Soda ash (3 gpl). ML ratio is 1:10. In vat dyeing the leveling agent used is Dispersol VL (1–2 gpl) (ICI). It is non-ionic in nature and helps regulate the initial strike rate of high affinity dyes. The higher concentration is required for pale shade and lower concentration for medium depth shade.

The vat dyeing process (method 1) is described as below.

1. At room temperature add Setamol WS (1 gpl) – dispersing agent and leveling agent dispersolvl (0.5 gpl) and run for 10 min.

2. Then add dye (already pasted with Setamol WS) using dosing method (dosing time 20 min).

3. Then raise temperature to 75°C at the rate of 2°C/min.

4. Hold at 75°C for 20 min.

5. Then cool the temp to 60°C.

6. After 10 min add half the quantity of caustic and run for more 10 min. It is better if caustic is added by dosing (time 10 min).

7. Next add half quantity of caustic soda and hydros with slow dosing. Run for 20 min.

8. If you are following method 2 then at this point add common salt dosing and run for 40 min (for blue shade use Glauber salt)

9. Then overflow drain.

10. Cold wash and rinsing for 10 min.

11. Oxidation is carried at 60°C for 20 min using Hydrogen Peroxide (3 gpl) and Acetic Acid (1 gpl).

12. Then cold wash and rinsing for 10 min.

13. Then do soaping using anionic detergent (1.5 gpl) at 95°C for 20 min. Use soda ash (1–2 gpl) only for soaping of any blue shades.

14. Two hot wash at 80°C for 15 min each.

15. Then add cationic softner (2 %) and run for 20 min at 55°C.

16. Unload, hydroextract and dry.

The above process is known as hot pigmentation process. Blue BC is sensitive to over reduction. Protect with sodium nitrite or glucose (2 gpl). Always use soft water since calcium and magnesium salts present in hard water causes precipitation of leucovat resulting into patchy dyeing weakening the strength and poor rubbing fastness in case of deep shades. Otherwise use adequate sequestering agent (sod. hexameta phosphate /sod salt of EDTA) should be incorporated in the bath. It is necessary to maintain dyebath in reduced state during dyeing. It may be noted that excess quantities of caustic soda and sodium hydrosulphite in dyebath reduces exhaustion while insufficient quantity causes oxidation of dye batch thereby causing patchy dyeing.

Therefore, presence of caustic soda in dyebath should be checked with phenolphthalein paper which turns red and presence of sodium hydrosulphite is checked using vat yellow paper which turns blue. Stripping of vat dyes can be done by using Dispersol VL (2–5 gpl) in blank reducing bath containing caustic soda and sodium hydrosulphite. Vat dyes are specially suited to meet the stringent fastness requirements of end use fabrics like.

1. Work wear and uniforms for defense, police, firefighting, postal and nursing services, which undergo severe laundry washing and long exposure to day light.

2. Clothing and household articles like office wear, sportswear, household linens, towels and bed sheets, which go in for frequent washing often followed by drying in direct sunlight.

3. Furnishings like curtains and seat covers needing high fastness to light.

4. Outdoor use fabrics like tenting and tarpaulins where fastness to weathering is essential.

5. Yarns for sewing threads, top dyed suiting and shirting requiring fastness to hypochlorite bleaching and severe washing.

Vat dyeing is having a very bright future. It will be used for fiber and yarn dyeing. It will pass all the fastness test required by the foreign buyers. Our technicians should all gear up for the job. They should remove all fear from their mind that vat dyeing is very difficult. Fabric dyers are all expert in vat dyeing. Vat dyes so named: the word vat means vessel. The dye takes their generic name from vatting, the vat dyes are naturally obtained coloring matter from the ancient time and kept into wooden vat and make solubilize in vat by the process of fermentation – so it is called vat dye.

Vat dyes are complex organic molecules which are insoluble in water, but when their carbonyl groups are properly reduced in a solution of caustic soda and sodium hydrosulphite to the so-called leuco or soluble state, they exhibit an affinity for cellulosic fiber. Chemical oxidizing agents are usually used to hasten oxidation of the reduced dye within the fiber back to its insoluble form which is physically trapped. This results in shades usually of excellent wash, chlorine and light fastness. The dyes are sold as powders or pastes which form dispersions upon dilution with water.

Despite their relatively high cost and difficult methods of applications, anthraquinone vat dyes are one of the most important groups of synthetic dyes because of their all-round superior fastness. The liquid types have become more important as the result of the progressive automation in textile dye houses and color kitchens. Improvements in the physical form of dyes have made towards rationalization of the dyeing process. The average particle size after a milling process is well below 1 um. The vatting rate is determined by the particle size distribution and crystalline form of dye.

5.2   Classification of vat dyes

Vat dyes can be classified based on the following:

(a) Chemical constitution

(b) Method of application

As per chemical constitution, vat dyes can be classified into following two classes:

1. Indigoid

2. Anthraquinonoid

Indigoid dyes have limited range of colors. These form of dyes can dye nylon, wool and cotton. Whereas Anthraquinonoid vat dyes, i.e.

anthraquinonoid compounds with carbonyl groups are in the market since the beginning of 20th century. Most of the vat dyes belong to anthraquinone groups because indogoid have lost their practical importance. All the browns, khakis, olives, grey products belong to this class.

5.2.1   Classification on the basis of application

Vat dyes are classified into four groups according to method of application, depending on the solubility, ease of vatting, optimum conditions of dyeing etc. In Germany vat dyes are divided into four groups, according to the method of application.

1. IK Class (K for cold or Kalit in German)
2. IW Class (W for warm)
3. IN Class (N for normal)
4. IN Spl Class (S for special dyes)

IK Class – This group consist of those dyes which are wetted and applied at lower temperature (20–30°C). These dyes require lower concentration of reducing agent (Sodium Hydrosulphite) and alkali (Sodium Hydroxide) but addition of salt can be use to accelerate complete exhaustion. These are known cold process dyes too. Certain yellow and orange dyes belong to this class of dyes.

IW Class – This type of dyes are applied at somewhat higher temperature, i.e. 40–50°C. Addition of salt is advisable for medium and dark shades. Yellow, Orange, Blue, Olive and Brown dyes are under this group.

IN Class – Almost all vat dyes belongs from this class are dyed at 60–65°C temperature. So high quantity of hydro sulphite and caustic soda is required. These dyes have excellent affinity to cellulose so no addition of common salt is required for exhaustion.

Few leveling and retarding agents are used to get uniform, level dyeing. Most of the Blue, Grey, Green, Red, Olive belong under this categories.

IN Special Class – Such class dyes need higher temperature 65–80°C and higher quantity of sodium hydrosulphite and caustic soda for perfect exhaustion. Addition of salt is not essential.

Navinon Grey 2B, Black CH, Black AC are such type of dyestuff.

5.7   Classification of vat dyes

For quinone vat dyes, there is no single classification according to dyeing properties, as is the case for the direct dyes. The German interessen

Gemeinschaft fur farbenindustrie (IG) developed one popular classification for their indanthrene range of vat dyes based on leuco compound substantivity and the required dyeing conditions.

Three main types:

1. The IN (indanthrene normal) group of dyeing temperatures (60°C) and dyeing temperatures (60°C). No salt is added to the dyebath because of the high substantivity of the leuco dyes for cotton.

2. The IW (indanthrene Warm) group of dyes requires the use of concentrated NaOH and lower vatting (50°C) and dyeing temperatures (50°C). The leuco forms of these dyes have moderate substantivity for cotton and some addition of salt is needed during dyeing to aid exhaustion.

3. The IK group of dyes only need a low concentration of NaOH with low vatting (40°C) and dyeing temperatures (20°C). These dyes have low substantivity for cotton and need considerable salt for good dyebath exhaustion. Some have amide groups that would be hydrolysed under the vatting and dyeing conditions used for IN and IW dyes.

There are special processes for some black vat dyes that require an oxidative after-treatment to develop the full black color. Table 17.1 compares the characteristics of these three types of vat dye. The required concentrations of hydros, caustic soda and salt increase with increasing amount of dye in the bath and with increasing liquor ratio.

There are various other classifications of vat dyeing methods. The SDC recommend tests to determine the best dyeing methods. In this, the color strengths of dyeing produced under different dyeing conditions are compared with those of standard dyeing using a grey scale. This test applies only to anthraquinone dyes. There are also SDC tests for determine the strike, migration and leveling characteristics of vat dyes. Different companies have different classification systems for third vat dyer. Because vatting and dyeing conditions vary from one another, the suppliers' recommendations should be consulted.

5.3   Vat dye application methods

In recent years there has been further development in dyeing of cellulosic fibers with vat colors. Vat dyes can be applied to cotton in almost any stage of manufacture, such as:

(a) *Batch-wise exhaust dyeing process* – Yarn in hank or package form, jigger dyeing, jet dyeing, soft flow machines are recommended for fabric dyeing.

(b) *Semi-continuous dyeing process* – It is well known that the usual method of vat dyeing in jigger by exhaust process is not satisfactory for thick fabric and also due to the consumption of higher amount of chemicals. It is therefore necessary to modify the method of application as below:

1. Pad-jig process (Pigmentation)
2. Vat acid process
3. Cold pad-batch process

(c) *Continuous dyeing process* – Various other suggestions for the continuous dyeing of cellulosic fabrics with vat dyes have been used. Continuous dyeing is economical particularly where long yardages have to be dyed in the same color. Following are the various applications:

1. Pad-steam process
2. Wet-steam process
3. Pad-dry-chemical pad process

5.4   Precautions in vat dyeing

To get uniform level dyeing results, the concentration of sodium hydro sulphite and caustic soda must be in exact quantity to reduce the dye bath during whole the process. Excessive quantity of reducing agent and alkali decrease the exhaustion rate of the dyeing. Similarly insufficient quantity of chemicals produce adverse effect on dyeing so to control this balance Phenolphthalien paper and Hydro papers are used.

(a) *Phenolphthalien Paper* – It is white paper which turns in pink color while contact with caustic soda.

(b) *Vat Yellow Paper* – It is gold yellow paper which turn in blue-violet color if hydro is available in dye bath.

It is also advisable that dyeing with two or three dyes combination must be from same group of dyes. So behavior of dyes is an important chapter to get uniform results. The leading vat dyes manufacturer are Atul, Meghmani, Indokem, etc. The leading vat dyeing units are Morarji Brembana, Nagpur, where they dye yarn and fabric using vat dyes. Nowadays, some Turkey units have started using vat dyes in India. Alok Industries, Wapi; Auro Dyeing, Baddi; Creative dyeing, Wapi, are also using vat dyes. Some big process houses in Bhilwara are doing job work for Digjam, Raymond, etc., using vat dyes for dyeing of the cotton or viscose part in the P/C or P/V blended fabric. Vat dyeing is the most difficult dyeing as compared to other dyeings. As on today there are few dyeing technicians in India who have mastered the art of vat dyeing.

5.4.1   Classification

Vat dyes are mainly divided into two main classes:

Indigoid vat dyes which are usually derivatives of indigotin (Fig 1, where R=NH) or thioindigo (Fig 1, where R=S)

Anthraquinoid vat dyes derived from anthraquinone (Fig 2)

The anthraquinoid group exhibits superior fastness properties – distinguished by excellent fastness to light and is the most widely used group. The derivatives of anthraquinone are more versatile and are made of higher condensed aromatic ring systems with a closed system of conjugated double bonds. The structural elements of several typical important classes of vat dyes are indanthrones, flavanthrones, pyranthrones, dibenzanthrones, isodibenzanthrones, benzanthrone acridones, anthraquinone carbazoles, anthraquinone oxazoles, etc.

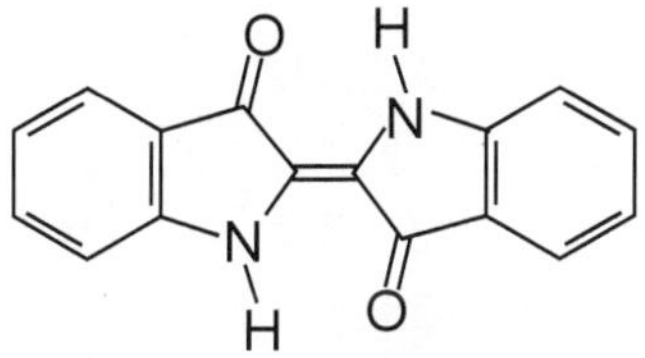

Indigo
C.I. pigment blue 66

Figure 5.1

The substitution of one and two –NH groups of indigo by –S group produces black, e.g. C.I. Vat Black 1, and brown, e.g., C.I. Vat Brown 5, dyes respectively. Substitution with two bromine and chlorine atoms on each of the two benzene rings of indigotin results in blue dyes, CI Vat Blues 5 and 41, respectively. While trying to find an anthraquinone analogue of indigo, indanthrone dyes were developed. They provide blue dyes. Since they are prone to over reduction and over-oxidation, they may create problem during dyeing. However, because of their attractive color, excellent fastness and moderate costs, they still form one of the most important classes of vat dyes. One important member of the class is Indanthrone Blue RSN, C.I. Vat Blue 4. Halogen substitution of indanthrone increases the range of shade. Thus C.I. Vat Blue 6 and C.I. Vat Blue 11 have two chlorine and two bromine atoms respectively at positions shown by * marks in Fig. 3. Approximately 80% of vat dyes belong to the anthraquinone chemical class of dyes and cover the full color range. One type, the *indigoid dyes*, includes indigo:

All vat dyes are insoluble in water. To apply them to a fiber, for example cotton, they are placed in an alkaline solution (Table 4). The insoluble dye is reduced to form a colorless (leuco) anion which is soluble and possesses affinity for the fiber. This is then adsorbed by the fiber, sometimes in the presence of sodium chloride, conditions similar to that for direct dyes. After the dyeing process the original insoluble parent dye is regenerated within the fiber by oxidation, usually using a solution of hydrogen peroxide or simply air:

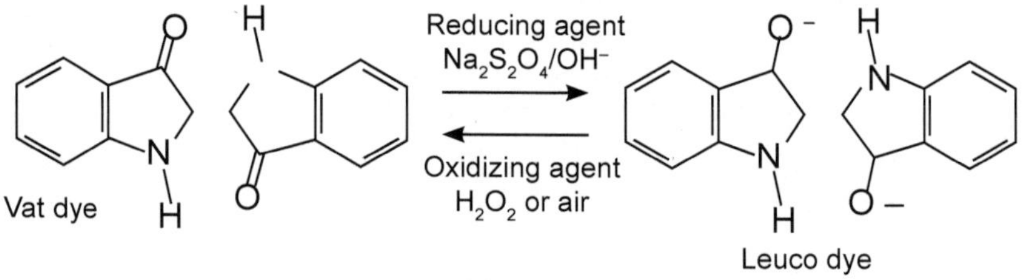

Figure 5.2

The dyes are insoluble within the fiber structure and therefore have good wash fastness and they also possess high light fastness.

5.5   Manufacturing of vat dyes

Flavanthrone, C.I. Vat Yellow 1 (Fig 4) was produced by the fusion of 2-amino anthraquinone at high temperature. In spite of slow oxidation, lower wash fastness and a tendency to photochromism, the dye is important for its good light fastness. This and its derivatives are in common with the indanthrones, among the oldest synthetic vat dyes.

A further development was the fusion of 2-methyl anthraquinone with caustic soda, then the dyestuff C.I. Vat Orange 9 or pyranthrone (Fig 5) containing pyrene structure is obtained. Dibromo and tribromo derivatives of the dye are C.I. Vat Orange 2 and 4, respectively. The dyes may cause fiber damage during dyeing and non-brominated product has inferior light fastness. However, they are used widely because of their good leveling properties, high color strength and moderate cost.

Another important dye intermediate is dibenzanthrone, from which many dark blues, navy blues, greens and blacks (CI Vat Blues 16, 19, 20 and

22, Greens 1, 2 and 9, and Black 9) vat dyes are derived. The best known being Indanthren Jade Green XBN, C.I. Vat Green 1 (Fig 6). They have disadvantages like lower fastness to rubbing, hot pressing and water spotting. Nevertheless, they have maintained their importance because of unique hues, good leveling and excellent fastness to wash and light. The isodibenzanthrone dyes include some interesting deep violet colors (C.I. Vat Violets 1 and 9) that have high color strength and good fastness to bleaching, but they suffer the same disadvantages as in the case of dibenzanthrone dyes. The benzanthrone acridones represent another large class, including C.I. Vat Greens 3 and 13, Black 25. They offer olive greens, olives, browns and grays. The leveling is difficult with these dyes, but they have excellent fastness and they protect fibers against the action of light.

Another extremely important class of dyes, the anthraquinone carbazole, includes C.I. Vat Oranges 11 and 15, Browns 13 and 44, Green 8 and Black 27. In spite of relatively flat colors, they have good leveling properties and excellent fastness to wash, light and chlorine.

The main representative of anthraquinone oxazole class is C.I. Vat Red 10, a strong brilliant red that levels well and has excellent fastness. The derivatives of perylene tetracarboxylic di-imide include C.I. Vat Reds 23 and 32, while those of imidazole include bright light fast C.I. Vat Yellow 46.

C.I. Vat Blue 66, having excellent leveling properties and good fastness to chlorine, is a derivative of triazinylaminoanthraquinone. Unlike indanthrone blue dyes, it is not susceptible to over-reduction.

5.6   Properties of vat dyes

1. Vat dye is water insoluble and can't be applied directly on textile material.
2. Mainly use fir cellulose fiber dyeing but in protein fiber dyeing PH should be controlled.
3. Rubbing fastness is not good.
4. Various shades are found.
5. Dyeing process is difficult.
6. Costly
7. Washing fastness of vat dye is very good with rating 4–5.

5.7   Problems with anthraquinone dyes

A number of chemical problems arose with some quinone vat dyes. These include:

1. Multiple reduction steps for poluquinones such as iudanthrone
2. Isomerism of leuco compounds to oxanthrones
3. Hydrolysis of amide groups
4. Over-oxidation after dyeing
5. Dehalogenation of some dyes

To minimize these types of problems, the supplier's recommendations for vatting and dyeing must be followed. Indanthrone (CI Vat blue 4) and some of its derivatives show a number of these problems. Indanthrone has two anthraquinone residues in its molecule. The normal blue leuco compound used in dyeing is that corresponding to the reduction of one of the anthraquinone groups (5, in figure 17.5). If both anthraquinone groups are reduced. The final product (6) gives a brownish yellow solution. Has poor substantivity for cotton and is more difficult to oxidize. Such over-reduction produces duller blue dyeing of lower color yield.

5.8   Vat dyes dyeing process

Vat dyes do not dissolve in water, while when reduced to be leuco salt by reducing agent under alkaline conditions, they can dissolve in water and get feature of immediacy with cellulose fibers, which is the way to achieve the purpose of dyeing. Then stable shade and good color fastness would come out via oxidation and soaping.

5.8.1   Reduction of dyes

Insoluble reducing dye will change into soluble leuco. Different levels of the best staining methods are brought out according to the reduction dyes' reduction potential and level of its immediacy to fibers.

5.8.2   Dye-uptake of leuco

The leuco is adsorbed by fiber and then it is diffusing into the fiber. It is necessary to use soft water or softened water in dyeing process. Hereby it is suggested to add the water softener for 1 g/L to some poor quality cotton fabric. The water softener would not only absorb calcium, magnesium ions, but also the iron, copper ions which are absorbed by the pipe into the dye bath, and finally keep the solubility of leuco.

The sodium sulfate could increase the dyeing absorption rate, and surely it is also a good promoter. So it can be added into the dye bathes with medium or low immediacy according to the demand. Referring to 3.13.

5.8.3   Leuco oxidizing

The leuco absorbed by fiber is oxygenated and changed into the former insoluble vat dyes, then it colors. Scour them in 2–3 baths with software before oxygenation, avoiding to be oxygenated with alkali.

Hydrogen peroxide is most suitable for suspension padding, while after oxygenation, it should be in soap boiling immediately without washing.

The sodium perborate is mild in very widely use. It can be bathed with the next procedure (referring to soap boiling).

5.8.4   Soaping treatment

The stable shade and good color fastness would come out via soap boiling.

Through high-temperature soap boiling, the floating color can be removed and the crystallization of vat dyes on fabric would be enhanced, so the fastness is increased, the stable and full color is achieved.

The soap lotion which is compounded by nonionic and containing chelating agents plays a better role in color light promoting.

5.9   Stripping of vat dyes

Vat dyes could strip colors from fabric or strip it into light color for dyeing again. General stripping materials contain the sodium hydrosulfite 5 g/L, peregal 2–4 g and 30% caustic soda 12–15 g/L. It should be operated in the temperature of 60–70°C for 30 minutes.

5.10   Shade rehandling methods

5.10.1   Washing method

It is the easiest way to amend the color light, which is suitable for the medium and dark varieties. If the fabric products are dark colored with more color floating or not fully washed, poor soaping effect, they can be amended by washing or soaping.

5.10.2   Reduction steaming

If the cloth is dyed to dark by vat dyes which need to be reduced to a light color, use the caustic soda and sodium hydrosulfite for 25 g/L each while in steaming washing, over 1/10 color light can be reduced.

5.10.3   Whitening agent rehandling method

It can be used to reduce the red color of the fabric dyed by the reduction vat dyes and improve the brightness of cloth, especially for the medium or light. General dosage of brightener is 0.3–1.2 g/L.

The method of stripping for vat dyes and sulfur dyes is same. In terms of sulfur dyes, it will consume more caustic soda and hydrosulphite. Please agitate evenly before using the stripper solution.

(1) Application for continuous process:

Recipe of padder solution:

KD-5A (Stripper): 20–40 g/l

Hydrosulphite (85%): 20–50 g/l

NaOH (100%): 20–50 g/l

Steam process of stripping is as same as the process that is applied for dyeing.

Notes: Adequate washing is necessary after steaming process. Suggest to add prepared padder solution (above) 15% together with water for first washing. Consumption amount of stripper, caustic soda and hydrosulphite depend on how deep the color is and how many dyestuff could be removed.

(2) Application for non-continuous process:

KD-5A (stripper): 2–6 g/l

Hydrosulphite (85%): 5–20 g/l

NaOH (100%): 5–20 g/l

Bath ratio: 1:10~~1:40 (Bath ratio is as big as possible)

Period for processing is 30–60 min at 60–80°C (The period and temperature depend on property of dyestuff which should be removed)

Suggestions: Add a certain amount of chelating dispersant given the amount of dyestuff to be stripped are more.

Significant results can be achieved when add DA-BS800 to stripping bath of vat dyes. While the stripping effect depends on the amount of DA-BS800, recommended dosage for stripping is as follows when in big liquid ratio: 3–5 g/L DA-BS800

As refer to deep color, suggest adding protective colloid, such as Dekol SN-S (2 g/L) or dispersing agents such as Setamol WS or Uniperol AC (2 g/L) to avoid precipitation. If it is hard water, we recommend adding Dekol SN-S or Trilon TB. The stripping temperature should be as high as possible, but cannot exceed 85°C in order to maintain the stability of the dye bath.

Recipe:

- 10–15 ml/L Sodium hydroxide
- 5–6 g/L Sodium hydrosulfite
- 3–5 g/L DA-BS800

Keep for 45–60 min at 60–85°C (checkout the reduction state), then warm washing, cold washing. The greater the dye bath, the smaller risk of dye precipitation in DA-BS800 solution.

Light color 1:5; Medium color 1:8 to 1:10

Deep color 1:10 or more greater

5.10.4 Discharge printing process

Dyeing → Discharging → Drying → Streaming → Washing (2 s) → Soaping (3–5 g/L Sodium carbonate, 90–100°C, 10 s) → Finishing

Vat dyes	X
Sodium formaldehyde sulfoxylate	10%
Urea	6%
Sodium carbonate	6%
Starch paste	Y

Chrome dyes (Mordant dyes)

6.1 Introduction

Chrome dyes were extensively used for the dyeing of wool up to First World War. Certain acid dyes are capable of combining with chromium within the fiber. Kostanecki was the first to use the term "mordant dyeing" to describe the formation of aluminum or iron complexes of alizarin or metal related hydroxyanthraquinones. The complexes are less soluble than the acid dyes thereby improving the wash fastness in wool dyeing. Acid dye Naphthol Blue Black became the basis for low-cost navy and black dyeings of wool and widely used. Despite good fastness and economic advantages, the residual chromium ions discharged into streams and other water sources has posed an unfavorable position in relation to ecological damage.

By middle of 1930s the development of chrome dyes was virtually completed. Structurally suitable dyes were developed for simultaneous dyeing and chroming process. In the 1960s, an interesting development has been the marketing, by Fran, of water-dispersible chrome dyes which only become soluble as the temperature of the dye bath reaches 70°C. In 1996, Ciba introduced a reactive dye Lanasol Black PV brands to replace chrome dyes Eriochrome Black PV. Although reactive dyes cannot yet replace chrome dyes, they are claimed to represent an interesting alternative in most cases.

From the dyer's point of view, chrome and pre-metalized dyes are considered as acid dyes. They can be applied on protein fibers requiring a chromium salt as mordant to obtain complete fixation on the fiber and hence mordant dyes for wool are usually called chrome dyes. In other words chrome dyes or acid mordant dyes are acid dyes having additional groups which enable the dye to form a stable co-ordination compound with chromium or with the fiber, thereby improving light and wet fastness.

Chrome dyes are not as bright as acid dyes and the shades are similar to 1:2 metal complex dyes. The latter dyes are suitable for light to medium shades, while chrome dyes are best for heavy dark shades. The light fastness of some chrome dyes is not good in light shades. The dyes of both dye-classes are often brightened by adding to the recipe a small amount of milling acid dyes. More dyes and labor are required in dyeing with mordant dyes. On the other hand, mordant dyes seldom give such bright shades, as do the acid dyes.

Despite numerous predictions of the demise of chrome mordant dyes over the last few years, it is clear that they are still important in the market place due to their extraordinary fastness properties and economy, especially on heavy dark shades like navy and black.

6.2   What are chrome dyes?

- Mainly used for black and navy shades on wool
- Still approx. 30% of all dyes used for wool are chrome dyes
- These dyes require the after treatment with a mordant to develop the fastness properties
- The mordant for chrome dyes is potassium dichromate (heavy-metal salt)
- Chrome dyes have a special position in wool dyeing, since when applied by the after chrome method, they have very good level-dyeing and migration properties and excellent washing fastness after chroming.

6.3   Classification of chrome dyes

Chrome dyes may belong to various chemical classes, namely:

1. Azo
2. Anthraquinoid
3. Triphenylmethane
4. Xanthene

(1) *Azo* – About 80% of chrome dyes belong to Azo group, mostly monoazo with a few important disazo dyes. The Azo chrome dyes cover the whole hue range except bright blues, violets and greens. Azo chrome dyes show highest fastness to light and wet treatments. They can be applied by all three mordanting methods described later, with a few exceptions. The Azo dyes may be subdivided into:

1. o,o'-dihydroxyazo dyes, the most important member of the sub-group is C.I. Mordant Black 11. Other important members belonging to this sub-group are C.I. Mordant Green 15, C.I. Mordant Black 15, and pyrazolone derivative, C.I. Mordant Red 7.
2. o-amino - o'- hydroxyazo dyes incorporate important brown dyes such as C.I. Mordant Brown 48. Other important members are C.I. Mordant Browns 1 and 33. These dyes have less aqueous solubility, especially in the presence of acid, due to the absence of sulphonic and carboxylic groups.

3. Salicylic acid derivatives incorporate nearly all yellows and oranges and a few of the browns, and are characterized by bright hues, e.g. C.I. Mordant Yellows 3 and 5.

The above three sub-groups include most of the Azo chrome dyes. Other smaller sub-groups containing important individual Azo chrome dyes are:

(a) o-hydroxy –o'-carboxyazo dyes incorporate one important dye, C.I. Mordant Red 9

(b) Azo chrome dyes in which the metal–complex formation is accompanied by oxidation to the quinine form, e.g. C.I. Mordant Black 9.

(2) *Anthraquinoid* – The parent dye in the anthraquinoid group is alizarin or 1,2-dihydroxyanthraquinone. The dye was originally obtained from madder, a natural product and then manufactured synthetically. It has no affinity for wool and can be applied after mordanting. The sulphonated product (3- sulphonic acid), C.I. Mordant Red 3 is water-soluble and may be applied by the after-chrome method. The other members, e.g. Anthracene Brown (1,2,3-trihydroxy anthraquinone) and Alizarin Orange (3-nitro derivative) have very low aqueous solubility. The most important chrome dye of this group is C.I. Mordant Black 13 and 38.

(3) *Triphenylmethane* – Triphenylmethane group incorporates important bright blue chrome dye which has moderate light fastness.

(4) *Xanthene* – Chrome dyes belonging to Xanthene group are very few, but brightest chrome red C.I. Mordant Red 27, belongs to this class.

6.4   Application methods

Chrome dyes are mostly dyed by after-chroming method, but can also be dyed on wool by pre-chromed and metachromed or directly together with the dye without any after-treatment. During application of chrome dyes, it is not necessary to fix the dye on wool, but also to apply the chrome mordant ensuring the combination of dye and mordant inside the fiber. Historically, dyers used to apply natural mordant dyes with different mordants in order to produce a range of colors from each dye. With the development of synthetic mordant dyes, however, chromium has become the almost universally used metal in mordant dyeing. The chromium salts are main mordant used due to cheapness and less variation in shade of dyeing. The chromium compound used is sodium or potassium dichromate. However, the method used depends on the class of dye used and type of material being dyed. There are following three methods for application of chrome dyes:

1. Chrome mordant method (pre-chrome method)

2. After-chrome method

3. Meta chrome method (chromate method)

(1) *Chrome mordant method* – It is a tedious process that needs two-bath process for its application. Moreover dyed shades have less fastness properties compared to those obtained by the after-chrome method or chromate method. The application involves pre-mordanting of wool with chromium salt (i.e. commercial name potassium dichromate or sodium bichromate) before dyeing application. Therefore this method is called 'pre-chrome method' too. This is the oldest but less important method than the other two. In this method wool is first mordanted with a chromium compound and then dyed. The method gives good coverage of wool fibers of different dye affinity and permits simple shade matching. Chrome mordant process requires two separate baths and is consequently expensive on time, energy and water. Fiber damage is also more as compared to modified after-chroming method. The wash fastness is also slightly lower than that obtained in the latter method. The method is particularly suitable for light and medium shades on worsted fabrics. Mordanting of woolen yarn is done using:

- Potassium dichromate or sodium bichromate – 1–3%

- Ammonium sulphate – 2% (pre-dissolved)

Dye bath is set at 60°C. Gradually, temperature is raised up to boil and mordanting is employed for 60–90 minutes. Then machine is drained. Then hot wash. The wool is taken for dyeing in another vessel. The dye bath is set at 50–60°C with pre-dissolved dyestuff. Raise the temp up to boil within 30 min and dye 60–90 min. For better milling fastness, a further treatment with 0.5% dichromate for 30 min at boiling may be recommended. Finally material is rinsed with hot water, soaped and cold washed. It is oldest method of dyeing as well as two-bath process which needs maximum water, labor, space and time tó dye one shade so this process is not in common use.

(2) *After-chrome method* – The method is widely used but it has disadvantages that the final shade cannot be judged unless complex formation is completed, and it is a two-stage process. The first stage is a simple acid dyeing followed by addition of dichromate to the exhausted dye-liquor and further boiling. In the second stage chroming of wool and formation of chromium-dye complex takes place in the fiber simultaneously. This method gives better fastness to milling and potting in dark shades. The dyeing and chroming processes, although separate steps, are often carried out in the same bath, thereby reducing dyeing time, water and energy requirements.

However, the dichromate should be added after exhaustion of the dye, otherwise the chromium dye complex may precipitate in the dye bath. Run the woolen material in the following bath containing:

- Acetic Acid (40%) – 3–5%
- Glauber salt – 5–10%
- Lyogen SMK – 1–2%
- Time required – 15–20 min
- Temp – 50–60°C

Then add pre-dissolved dyes in two installments with an interval of 15 minutes. Raise the temperature up to boil in 30–40 min and dye for 45–60 min. To get better exhaustion if required 1–2% formic acid is added. Check if the dye bath is absolutely exhausted and liquor is colorless. Then cool the dye bath up to 70°C.

To get better leveling, acetic acid is replaced by 5% Ammonium sulphate or Ammonium acetate. Now add Potassium dichromate (1–2%) and raise temperature up to boil within 30 min.

Material is chromed here for 30–45 min at boil. Then drain, hot wash, soaping and cold wash. Regarding concentration of chromium salt, it should be half of the dyestuff weight. But application of potassium dichromate must not be more than 1.5%. After complete exhaustion of dye, potassium dichromate should be added otherwise unlevel and patchy dyeing will occur.

(3) *Metachrome method* – In this method mordanting and dyeing is done simultaneously by controlling the conditions properly and selecting suitable dyes. Almost all Azo chrome dyes are suitable for metachrome process of dyeing, except for some marine blues and blacks. The reaction takes place simultaneously in single-bath chrome dyeing. This is a popular process of dyeing because of its simple dyeing application. Final desired shade can be matched during dyeing. Handle of dyed material seems better compared to dye with rest methods. This is a single-bath process which shortens the dyeing process to achieve cost economy by all the means. Dye bath is set with pre-dissolved dyes at 50–60°C. Run the material for 15 minutes. Add the following:

(a) Potassium dichromate – 1–2%

(b) Ammonium sulphate – 2–5% (pre-dissolved)

Raise the temperature up to boil and dye for 60–90 min

Especially for medium to dark shades, formic acid 0.25–0.5% is added to get better exhaustion.

Following are the chrome dyes used today which is manufactured by Clariant.

(a) Omega chrome Black T Supra

(b) Omega chrome Yellow KI

(c) Omega chrome Bordeaux BRI

(d) Omega chrome Fast Blue BI

(e) Omega chrome Brown 2RI

Following are the disadvantages of chrome dyes:

1. Long dyeing time

2. More fiber damage

3. Severe changes in color during chroming

4. Difficulties in correction of faulty dyeing

5. Chromium residues in the effluent

The above disadvantages may eventually lead to ceasing the use of chrome dyes in near future.

Before the advent of synthetic dyes, all dyes came from natural sources such as minerals and plants. Often these dyes faded quickly if the dyed material was laundered. To fix or stabilize the color, chemical agents called mordants were used. Chemically, the mordant binds with the dye and the fibers of the material, preventing bleeding and fading. As early as 1820, the cotton and wool industries were using large amounts of chromium compounds (such as potassium bichromate) in the dyeing process. Red and green pigments developed from chromium compounds were also used for printing wallpaper during this period.

In 1822, a man named Andreas Kurtz moved to England and began producing potassium bichromate and selling it to the English textile industry at 5 shillings a pound. Competition soon drove the price down to 8 pence, about an eighth of the original price. This did not give Kurtz a satisfactory profit, so he began producing other chrome compounds, specifically chrome pigments. His chrome yellow achieved cult status when Princess Charlotte, daughter of George IV, used it to paint her carriage. This was perhaps the origin of the "yellow cab," an idea exemplified today in New York City taxis. Kurtz left his mark on the world of color; "Kurtz yellow" is still available in British color catalogues.

In the film Erin Brockovich (2001, Universal Studios) Pacific Gas and Electric is portrayed as a corporate giant that poisoned the water of the small town of Hinckley, California. The movie, which is based on a real lawsuit, suggests that high levels of chromium-6 in the groundwater were responsible for an eclectic range of diseases among residents there, including various cancers, miscarriages, Hodgkin's disease and nosebleeds. In 2010, the Environmental Working Group studied the drinking water in 35 American cities. The study was the first nationwide analysis measuring the presence of the chemical in US water systems. The study found measurable hexavalent chromium in the tap water of 31 of the cities sampled, with Norman, Oklahoma, at the top of list; 25 cities had levels that exceeded California's proposed limit of Chromium VI and its less toxic forms.

Mainly used for black and navy shades on wool, still approx. 30% of all dyes used for wool are chrome dyes. These dyes require the after treatment with a mordant to develop the fastness properties. The mordant for chrome dyes is potassium dichromate (heavy-metal salt). There are very few synthetic dyestuffs currently in use that require a separate mordant, except for some dyes for wool, where mordant dyes are still quite popular. Since chromium is almost exclusively used as the mordant on wool, chrome dye has become essentially synonymous with mordant dye. Many natural dyes (plant extracts, etc.) require a mordant. The mordant used can significantly influence the hue produced with a particular dyestuff. These are special acid dyes in which certain metal atom can be introduced during dyeing. These are water-soluble dyes and affinity for silk, wool and polyamides. Mordant dyes require a mordant in their application and these dyes upon combination with the mordant deposit on the fiber in the form of insoluble color. Most commonly dyes have hydroxyl or carboxyl groups and are negatively charged (anionic) in nature.

6.5   Properties of mordant dyes

1. These dyes are economical dyes and are generally used to produce dark shades such as dark greens, dark blues and blacks.

2. These dyes have good leveling and color fastness properties.

3. The interaction between fiber and dye is established through very strong ionic bonds, which are formed between the anionic groups of the colorant and ammonium cations on the fiber. Chromium or the metal ion acts as bridge between the dye and fiber, which gives rise to a very strong linkage, resulting into excellent fastness properties.

However there are disadvantages of the chrome dyes also such as longer dyeing cycles, difficulties in shading, risk of chemical damage to the fiber and the potential release of chromium in the wastewater.

6.6   Mechanism of dyeing

These are a special class of acid dyes, which are soluble in water and applied to the fiber from an acidic bath. When a solution of an acid mordant dye is mixed with a solution of potassium dichromate in the presence of sulfuric acid, chromium ion from dichromate forms a complex with the dyes, this complex is insoluble in water, and hence precipitates on the fiber.

6.7   Application of mordant dyes

There are three different methods of application of chrome dyes on the fiber:

1. Chrome-mordant method
2. After-chrome method
3. Meta chrome method

(1) *Chrome-mordant process* – In chrome mordant process, the fiber is first treated with potassium dichromate in the neutral bath or in the presence of either sulfuric or formic or oxalic acid. When sufficient amount of chromium is taken up by fiber, it is taken out, squeezed and entered in the dye bath containing acid mordant dye. The dye form an insoluble complex with chrome present on the fiber.

(2) *After-chrome method* – In this method the substrate is first treated with the dye, the dye is exhausted by the addition of an acid, and after complete exhaustion the material is taken out squeezed and then run in a solution containing potassium dichromate and an acid. Metal dye complex is formed on the fiber, which is insoluble.

(3) *Meta chrome process* – This is a single bath process, in which the material is treated with in a bath containing acid mordant dye, potassium dichromate and ammonium sulphate. The dye along with potassium dichromate and ammonium sulphate got absorbed by the fiber and evenly distributed but no complex is formed because the pH is not suitable for the chemical reaction to take place.

In the second step of the meta chrome process, when the dye bath is heated, ammonium sulphate is converted into ammonia and sulphuric acid, which makes the bath strongly acidic and potassium dichromate in the presence of strong acid now react with the dye molecule forming an insoluble complex on the fiber.

A dyeing cycle for dyeing of wool with chrome dyes is shown below:

Here,

A = Gluber salt + Acetic acid + Leveling agent

B = Dye
C = Potassium Dichromate
D = Acid

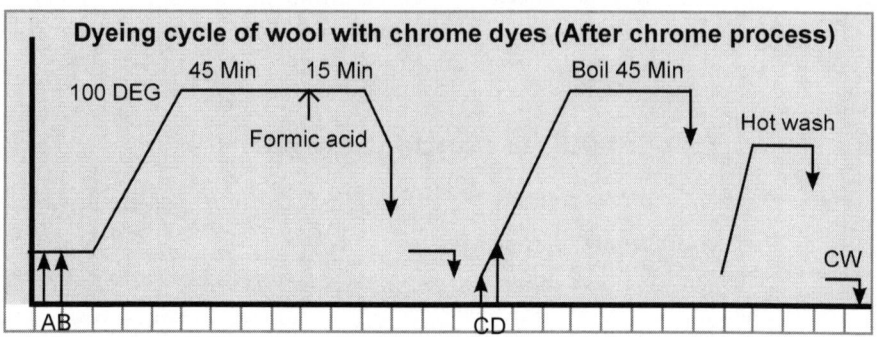

Dyeing cycle of wool
Figure 6.1

Applicability

Mordant dyestuffs are generally used for protein (wool and silk). They are practically no longer used for polyamide fibers or for printing

Properties

Thanks to their good leveling properties and very good wet fastness after chroming, chrome dyes are used principally to obtain dark shades (greens, blues and blacks) at moderate cost. There are disadvantages, however, in their use: long dyeing times, difficulties with shading, the risk of chemical damage to the fiber during chroming and the potential release of chromium in wastewater.

Chemical characteristics and general application conditions

The *Colour Index* classifies these colorants as mordant dyes, but chromium has become the almost universally used mordant and the class is commonly referred to as chrome dyes.

From a chemical point of view they can be regarded as acid dyestuffs that contain suitable functional groups capable of forming metal complexes with chrome. They do not contain chrome in their molecule, which instead is added as dichromate, or chromate salt to allow dye fixation.

Interaction with the fiber is established through ionic bonds formed between the anionic groups of the colorant and ammonium cations available on the fiber. In addition chromium acts as a link between dye and fiber. This gives rise to a very strong bond, which is reflected in the excellent fastness obtained. Figure 6.2 shows the ionic and coordination bonds in the case of wool.

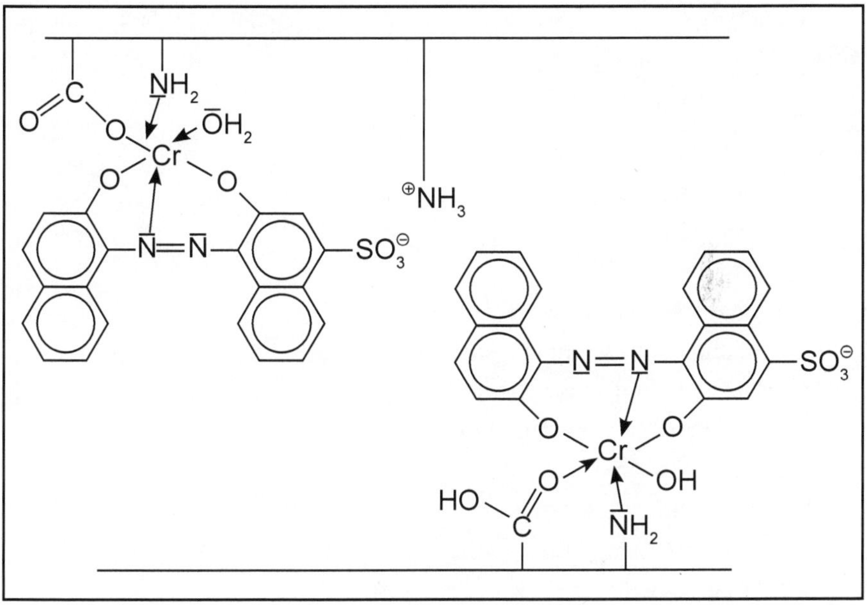

Figure 6.2

6.8　　　What are the risks with chrome dyes?

- Mistakes in wrong handling of the process in the dye house
- Endangers health and even the life of workers
- Causes serious water pollution
- Pushes extractable heavy metal in garments above all limits
- Less spinning yield due to fiber damage
- Harsh wool handle
- Negative image for brands

6.9　　　Replacement of chrome dyes – Why?

- Many brands have heavily restricted the use of chrome dyes
- Scandinavia wool industry: totally metal-free

- Effluent restrictions
- Health care for workers
- Better spinning yield and wool quality
- An industrial scale comparison

In 1997, the first LANASOL CE dyes were launched to initiate the replacement of chrome dyes

Reliable tests:
- One of the largest worsted mills in the world
- 2 tons of wool
- One dyed with chrome black
- One with LANASOL black
- 19.5 micron – ecru wool
- Convincing commercial and sustainable results
- Color fastness and shades depth
- Preservation of the natural bulkiness
- Preservation of fiber quality
- Performance in spinning
- Performance of the yarn
- Significant cost savings due to less waste
- More sustainable

Metal-complex dyes

7.1 Introduction

During the last half of the 19th and early part of the 20th century, the use of chrome mordant dyes on wool was widespread due to excellent all round fastness properties. However, several disadvantages of the dye-class, such as complicated and lengthy processes leading to damage of wool, change of shade after chroming and difficulty in shade matching, were well known. As a consequence, a much simpler dye-class, metal-complex or pre-metallised dyes, was introduced in the early 20th century. Acid dyeing, 1:1 pre-metallised dye was introduced in 1919 and neutral dyeing 1:2 pre-metallised dyes in 1951. The metal-complex dyes are also pre-metallised dyes. The complexes are co-ordination compound and a co-ordinate link is formed between the metal and dyes. As the name implies, in metal-complex dyes one metal atom, commonly chromium, is complexed with either one (1:1) or two (1:2) molecules of a typically monoazo dye that contains groups (such as hydroxyl, carboxyl or amino) that are capable of coordinating with the metal. In the Colour Index, metal-complex dyes are included in the acid dye class, though they have distinct differences from non-metallised acid dyes.

Metal-complexed dyes are closely related to chrome mordant dyes; but from dyers point of view they may be considered as acid dyes. These dyes were originally developed to overcome the disadvantages of chrome dyeing of wool. Metal-complex dyes are applied on both wool and synthetic polyamides because of their superior colourfastness to wet treatment. The metal-complex dyes are brighter than chrome dyes but duller than acid dyes. The metal-complex dyes also face pressures of ecological problems similar to those affecting chrome dyes. Many large mills have already banned their use and consequently it is becoming difficult to obtain muted shades of the required light fastness and levelness especially on nylon. Reactive dyes may be used with good wash fastness properties, but adequate light fastness is difficult to achieve.

The 1:1 and 1:2 metal-complex dyes resemble leveling and milling nonmetallised acid dyes respectively in terms of general application conditions. The increased popularity of these dyes, which has occurred mainly at the expense of mordant dyes, can be attributed to health and effluent hazards associated with the use of chromium in mordant dyeing and the comparatively simple application procedures of metal-complex dyes. The 1:1 metal-complex dyes have been gradually replaced by 1:2 metal-complexes due to superior all round fastness properties and less damaging application conditions of the latter.

As a logical extension of the use of metallization of chrome dyes on the fiber, it thought that if chromium is incorporated in the dyestuffs molecule by the dyestuffs manufacturer, the pre-metallised dye can then be applied in the fiber by a simple process. This would eliminate chroming process by the dyers. This was achieved around 1915 independently by Ciba Ltd and IG Farben Industries AG who produced Neolan dyes and Palatine fast dyes, respectively.

Due to their ease of application and fastness properties, they are widely used for dyeing tropical suiting, high-class dress materials, hosiery yarn and carpet yarn. In the decade of 1940–50, a number of 1:1 metal-complex dyes had firmly been established in the wool dyeing trade. In 1951 Geigy developed a new series of dyes under the trade name 'Irgalan' 1:2 metal-complex dyes. In 1970 BASF introduced Acidol M dyes. Now dyes are available in market which contain metal complex (chromium or cobalt) in their structure.

Now dyes are available in market which contain metal complex (chromium or cobalt) in their structure. These dyes need no metal chemical during their application. These dyes are equally brighter as acid dyes but fastness properties (washing, staining, seawater, light, crocking and other wet fastness) are much excellent comparatively. In 1915 dyes manufacturer developed this type of dye in which metal was introduced. These dyes eliminate the chroming process for animal fiber if high degree of fastness is desired.

7.2   Classification of metal-complex dyes

Structurally metal-complex dyes are of two types:

(a) 1: 1 metal-complex dyes

(b) 1: 2 metal-complex dyes

Generally metal ions used is trivalent and hexa-co-ordinate and Cr 3+ and Co 3+ are both used, although chromium is most widely used. The 1:1 metal-complex dyes are widely used for dyeing loose stock and yarn dyeing for end uses such as floor covering, hand-knitting yarns and piece goods. They have excellent level dyeing properties and have good to very good color fastness to light. The 2:1 metal-complex dyes can be applied to wool at all stages of manufacture and they provide high colourfastness to wet treatments. The 2:1 metal-complex dyes are also important in union dyeings. The use of 1:1

metal-complex dyes has declined, but the use of 2:1 metal-complex dyes has increased because of their superior wet fastness properties.

There are two types of metal-complex dyes:

(a) Metal-complex (1:1) dyes (one molecule of an Azo group and one atom of metal compound). Generally metal ions used is trivalent and hexa-coordinate, and Cr 3+ and Co 3+ are both used, although chromium is most widely used. The 1:1 metal-complex dyes are widely used for dyeing loose stock and yarn dyeing for end uses such as floor covering, hand-knitting yarns and piece goods. They have excellent level dyeing properties and have well to very good colourfastness to light. The 1:1 metal-complex dyes was introduced in 1919. These dyes are sold under the trade names of Neolan (Ciba-Giegy), Palatine Fast (BASF), Ultralan (ICI), etc.

(b) Metal-complex (1:2) dyes (two molecule of an azo group and one atom of metal compound). These dyes have two dye molecules per metal atom. The first water-soluble 1:2 chromium complex dye for wool was introduced by Geigy AG in 1949 under the trade name of Irgalan dyes. The 1:2 metal-complex dyes can be applied to wool at all stages of manufacture and they provide high colourfastness to wet treatments. The 1:2 metal-complex dyes are also important in union dyeings. The use of 1:1 metal-complex dyes has declined, but the use of 1:2 metal-complex dyes has increased because of their superior wet fastness properties.

Two types of pre-metallised dyes are currently available: 1:1 pre-metallised type which is acid dyeing, i.e. they are applied from an acid dyebath at a pH of 2.0–2.5 using sulphuric acid. At this pH they have excellent migration properties but the equilibrium exhaustion is reduced. Under weakly acid conditions they have high substantivity but low migration properties. 1:1 metal-complex dyes are often used for the dyeing of woolen piece goods where good wet fastness is required. The very low pHs involved and the boil can cause the fiber to become brittle which can result in lower than average rubbing performance. 1:2 metal-complex dyes are more widely used because of their more favorable dyebath conditions being normally dyed at pH 5.5–7.0.

Three types of 1:2 pre-metallised dyes are available:

(1) Nonionic solubilize i.e. they contain no polar sulphonic acid groups, because they are weakly polar. They have excellent level-dyeing properties and cover wool of variable surface characteristics well. They are particularly useful for the dyeing of pale shades where high light fastness is required. They have high substantivity under weakly acid or neutral conditions and absorbed uniformly, with good coverage of tippy wool.

(2)   1:2 metal-complex dyes solubilised with one sulphonic acid group.

(3)   1:2 metal-complex dyes solubilised with two sulphonic acid groups.

The strongly polar 1:2 metal-complex dyes containing sulphonic acid groups are more sensitive to fiber surface variations such as sun damage to the fiber tips. As depth increases, however fiber selectivity decreases until with dark browns and navies it ceases to be a problem. The sulphonic acid solubilised dyes are more commonly used for the dyeing of dark colors.

Metal-complex dyes are pre-metallised dyes that show great affinity towards protein fibers. In this dye one or two dye molecules are coordinated with a metal ion. The dye molecule is typically a monoazo structure containing additional groups such as hydroxyl, carboxyl or amino, which are capable of forming a strong co-ordination complexes with transition metal ions such as chromium, cobalt, nickel and copper.

The chemical types are azo and anthraquinone giving a complete colour range. However they are duller than the acid dyes because of the presence within the dye structure of a metal atom. Chromium salts are often used although cobalt and nickel salts are also favored.

The metal atom forms a coordination complex with two molecules of a monoazo compound containing hydroxyl, carboxyl or amino groups in the 2,2' positions relative to the azo group. These compounds are called '1:2 metal-complex' dyes. An example is C.I. Acid Violet 78:

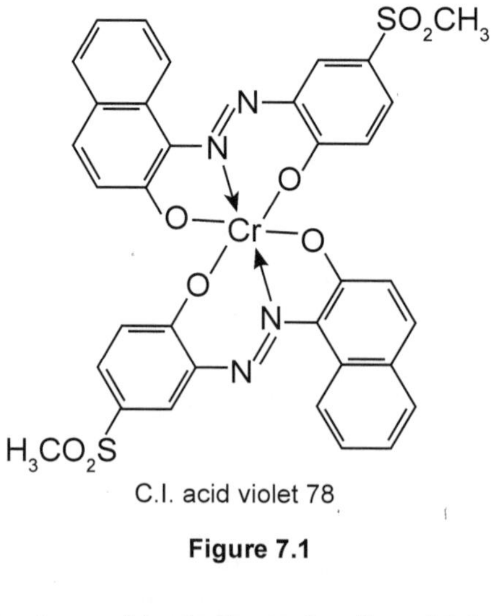

C.I. acid violet 78

Figure 7.1

Their application to wool is similar to that for acid dyes, but the pH value is restricted to the range of 4.5 to 6.0 (Table 7.1).

Table 7.1 Shows a comparison between 1:1 metal-complex and 1:2 metal-complex dyes.

Dye type	Leveling ability	Wash fastness	pH range
1:1 metal-complex	Good	Good	2
1:2 metal-complex	Poor	Very good	6–7

Metal-complex dyes belong to numerous application classes of dyes. For example, they are found among direct, acid, and reactive dyes. When applied in the dyeing processes, metal-complex dyes are used in pH conditions that are regulated by user class and the type of fiber type (wool, polyamide, etc). The pH levels for wool typically range from:

Strongly acidic (ranging from 1.8 to 4 for 1:1 metal-complex dyes)

Moderately acidic neutral (ranging from 4 to 7 for 1:2 metal-complex dyes)

7.3   Classification of metal-complex dyes

The classification of metal-complex dyes is based on the basis of number of dye molecules which are complexed with one metal ion.

1:1 metal-complex dyes: In 1:1 metal-complex dyes one metal ion is complexed with one dye molecule.

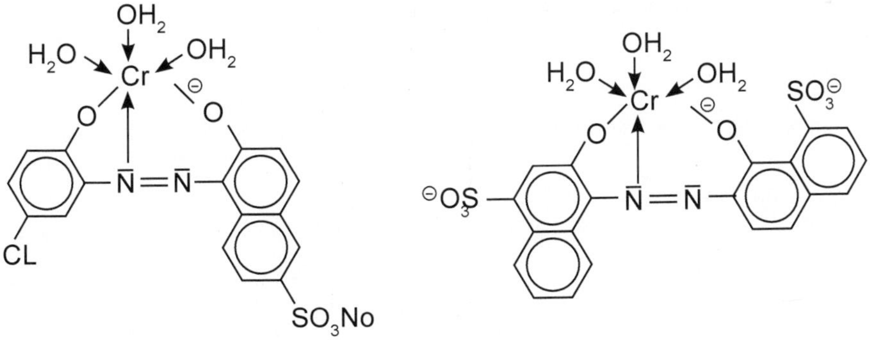

Acid violet 56
C.I. 16055

Acid blue 1586
C.I. 14880

Figure 7.2

1:2 metal-complex dyes: In 1:2 metal-complex dyes there are two dye molecules which are complexed with one metal atom.

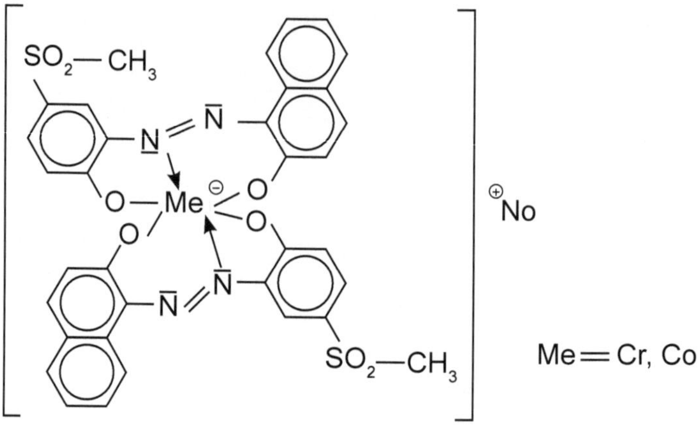

Figure 7.3

7.4   Features of metal-complex dyes

- Medium washing fastness
- Excellent, light-fastness
- Shows very good level dyeing and penetration characteristics
- Can cover up for the irregularities in the substrates

7.5   Types of metal-complex dyes

1. Non-sulphonated metal-complex dyes with high build-up capacity, low pH dependency and good wet fastness properties.

2. Monosulphonated metal-complex dyes with good build-up capacity, average pH dependency and good wet fastness properties. The ternary has a perfect combinability.

3. Disulphonated metal-complex dyes with highest wet fastness properties. Due to their high efficiency this range is especially suitable for dark shades. They have a high pH-dependence and are mainly used as a single dye or in a combination of two. The build-up is partially limited.

4. Ternary of perfectly coordinated metal-complex dyes with the highest light and wet fastness levels as well as an excellent, regular build-up. These ternary elements can be combined with selected Acid dyes. Due to the low electrolyte content the ternary is also suitable for printing carpets with digital spray processes; fast fixation is also guaranteed in a saturated steam medium.

7.6   Manufacturing of metal-complex dyes

(1) 1:1 Metal-complex dyes

Mono azo dyes (parent acid dye) are heated at temperature up to 130°C with an aqueous solution of a salt of trivalent chromium such as chromium (III) formate or fluoride, the pH being below 4.0. These dyes are soluble 1:1 chromium complexes of azo dyes and contain one atom of co-ordinated chromium atom per dye molecule with one or two sulphonic acid groups. The 1:2 complex is also prepared in a similar manner, but under alkaline conditions. The 1:1 metal-complex dyes include o-o'-dihydroxy, o-amino –o' – hydroxyl azo and derivatives of salicyclic acid. Thus, the ionic characteristics of 1:1 metal-complex dyes depend on the presence and quantity of sulphonic groups in the chromophore. Apart from chromium no other complexing metal has become of practical importance for this class of dye. It is possible to get a range of colours from yellow to black depending on the choice of diazo and coupling components. Chroming in organic solvents using a very high concentration of reactants and a very short reaction time is reported for ecological and economic reasons.

Solubility in water of 1:1 metal-complex dyes is conferred by the presence of one or more sulphonic acid groups, although C.I. Acid Orange 76 contains nonionic amino-sulphone or sulphamoyl groups ($-SO_2NH_2$) as solubilising aids. Such dyes containing no ionic solubilising group have an overall positive charge because of the presence of chromium cation. Depending on the nature and number of the solubilising groups and the nature of the monodentate ligands present, the conventional 1:1 metal-complex dyes either are effectively uncharged or carry an overall negative charge as in C.I. Acid Green 12 and Red 183, respectively.

(2) 1:2 Metal-complex dyes

These dyes have two dye molecules per metal atom. The first water-soluble 1:2 chromium complex dye for wool was introduced by Geigy AG in 1949 under the trade name of Irgalan dyes. The 1:2 chromium complexes are usually prepared in aqueous medium or in organic solvents using chromium (III) salts, in a weakly alkaline medium and or in presence of tartaric acid, lactic acid, citric acid and glycolic acid by the so-called mixed micellisation method. Formation of 1:2 cobalt complexes is used in few cases. Cobalt complexes absorb light in shorter wavelength than the corresponding chromium complexes and thus have higher fastness to light and lower rate of dyeing. Pure 1:2 chromium mixed complex can also be prepared from the corresponding 1:1 chromium complexes by the addition method. In the addition method, one mole of dye is converted into corresponding stable 1:1 chromium complex, isolated and then reacted with a further mole of same dye or a different one to give symmetrical or unsymmetrical 1:2 chromium complex.

In spite of lower wet fastness properties as compared to chrome dyes, 1:1 metal-complex dyes were popular for over a quarter of a century for their excellent migration and penetration properties, ease of application, good light fastness and comparatively brighter shades. However, their popularity diminished with the invention of neutral dyeing 1:2 metal-complex dyes in 1951 because of their applicability under weakly acidic to neutral conditions as against strong acidic conditions for 1:1 metal-complex dyes.

The neutral dyeing 1:2 metal-complex dyes are chromium or cobalt complexes of azo dyes. Like 1:1 complexes, they are prepared from azo dyes, in particular from oo'-dihydroxyazo dyes. A few of them are azamethine derivatives in which the azo group is replaced by –CH=N–, the nitrogen of which can form a coordinate bond with metals. Most of them are chromium complexes. A few of them are cobalt complexes, which have higher fastness to light and a lower rate of dyeing. The two parent dye molecules forming complex with each metal ion may be same or different, the symmetrical complexes being more stable. There are now four distinct types of 1:2 metal-complex dyes:

1. Weakly polar ones devoid of sulphonic groups in the dye molecule (structure III).These dyes are sold under the trade names of Irgalan (Ciba Geigy). Lanasyn (Sandoz), Isolan (Bayer), Ortolon (BASF), Capracryl (Dupont), etc.

2. Symmetrical 1:2 complex solubilised by two carboxyl groups per mole. Typical example of this class is Elbelan dyes marketed by L.B. Holiday.

3. Mono-sulphonated 1:2 metal-complex dyes of unsymmetrical type. Dyes of these types are Lanacron S (CGY), Lanasyn S (Clariant), Neutrichrome S (Fran), etc.

4. Di-sulphonated 1:2 metal-complex dyes with symmetrical molecular structure. The typical examples are Acidol M (Dystar), Azarine (HOE), etc. and contain two sulphonate groups.

The dyes with solubilising groups have proved to be very popular because they show high tinctorial yield, better fastness to wet treatments in full depths and cheaper preparation. The large number of patents published so far indicates that the structural model indicating by the classical 1:2 Cr and Co complex dyes require elaborate discussion in the field.

7.7   Application of metal-complex dyes

Metal-complex dyes is using for a variety of applications like wood stains, leather finishing, stationery printing inks, inks, coloring for metals, plastic, etc. As this dye is classified into two categories and both have different applications.

1:1 metal-complex dyes – These dyes have good leveling and penetration properties and are particularly suitable for application on carbonized wool. These dyes are applied under a strongly acidic bath at a pH of 1.8–2.5 with sulfuric acid or at a pH of 3–4 with formic acid, therefore these are not suitable for the blends having cotton component. Glauber salt is used as exhausting agent.

1:2 metal-complex dyes – These dyes are subdivided into two subgroups based on the solubilising groups present in the dye molecule. These dyes show moderate migration properties on nylon but very good overall fastness properties, because metal-complex dyes not only attach to nylon with ionic linkages, but also with coordinate bonds. The two subgroups are:

- Weakly polar dyes
- Strongly polar dyes

7.8   Dyeing with metal-complex (1: 1) acid dyes

In these dyes chemically one atom of chromium or cobalt is associated with one molecule of a monoazo dye (parent acid dye). Thus, the ionic characteristics of 1: 1 metal-complex dyes depend on the presence and quantity of sulphonic groups in the chromophore. Apart from chromium no other complexing metal has become of practical importance for this class of dye. It is possible to get a range of colors from yellow to black depending on the choice of diazo and coupling components. Chroming in organic solvents using a very high concentration of reactants and a very short reaction time is reported for ecological and economic reasons.

7.8.1   Dyeing method

Dyebath is set with predissolved dyes at 50°C temperature. Then goods are run for 15 min. Then add sulphuric acid in two installments with 5 min interval. It is advisable to use sulphuric acid at pH 2.0–3.0 for proper exhaustion and absorption. Run for10 min. Raise the temperature up to boil within 30 min and dyeing is carried out for 60–90 min at boil depending on the shade depth. Levelling agent can be used to get uniform dyeing. Glauber salt is also added to improve exhaustion of the dye. Ammonium sulphate or ammonium acetate is added as buffer to maintain the pH. Finally hot wash, then soaping and cold wash.

7.8.2   Merits of this class of dyes

1. Simple dyeing method
2. Uniform and level dyeing
3. Suitable for wool fibre, tops, hanks and fabric
4. Shading and brightening is possible
5. Excellent fastness properties such as washing, sea water, light, dry cleaning, crocking, chemicking, etc.
6. Very good reproducibility
7. Excellent dyeing behaviour in combination process
8. Good migration property

7.8.3   Drawbacks of this class of dyes

1. They have poor fastness to milling as compared to chrome dyes.
2. Addition of sulphuric acid must be done in 2–3 installments to avoid fibre degradation.

Following are the C.I. number of 1:1 metal-complex dyes used in textile units.

(a) Acid Yellow 99
(b) Acid Orange 74
(c) Acid Red 183
(d) Acid Red 184
(e) Acid Red 186
(f) Acid Red 194
(g) Acid Red 195
(h) Acid Blue 158
(i) Acid Black 52

7.9   Dyeing with metal-complex 1:2 acid dyes

In spite of lower wash fastness properties as compared to chrome dyes, 1:1 metal-complex dyes were popular for over a quarter of a century for their excellent migration and penetration properties, ease of application, good light fastness and comparatively brighter shades. However, their popularity diminished with the invention of neutral-dyeing 1:2 metal-complex dyes in 1951 because of their applicability under weakly acidic to neutral conditions as against strong acidic conditions for 1:1 metal-complex dyes.

The neutral-dyeing 1:2 metal-complex dyes are chromium or cobalt complexes of azo dyes.

There are now four distinct types of 2:1 metal-complex dyes.

(1) Weakly polar dyes (non-sulphonate groups). These dyes are sold under the trade name of Irgalan (Ciba), Lanasyn (Sandoz), Isolan (Bayer), Ortolan (BASF), Capracryl (Dupont).

(2) Symmetrical 2:1 complex soublised by two carboxyl groups per mole. Typical example of this class is Elbelan dyes marketed by L.B. Holiday.

(3) Mono-sulphonated 2:1 metal-complex dyes of unsymmetrical type. Dyes of these types are Lanasyn S (Clariant).

(4) Di-sulphonated 2:1 metal-complex dyes with symmetrical molecular structure. Dyes are Acidol M (DyStar). They contain two sulphonate groups. The dyes with solubilising groups have proved to be very popular because they show high tinctorial yield, better fastness to wet treatments in full depths and cheaper preparation.

7.9.1   Dyeing method

The dyebath is set with pre-dissolved dyestuff at 50°C temperature.

Following are the chemicals added:

- Ammonium acetate or Ammonium sulphate 2–4%, pH 5.5 to 6.0 with acetic acid.
- Dyeing is started for 15–20 min. Then add Glauber salt 5–10% (it acts as a retarding agent).
- Levelling agent 1–2%.
- Temperature is raised slowly up to boil and dyeing is carried out for 45–60 minutes at boil. To increase exhaustion further 1–2% acetic acid or 1–2% formic acid is added. Then hot wash, soaping and cold wash.

7.9.2   Merits of these dyes

1. Excellent fastness properties
2. Better level dyeing
3. Better reproducibility
4. Less damage on wool
5. Shorter dyeing cycle
6. More exhaustion of dyes

Following are the C.I number of 1:2 metal-complex dyes used in textile units.

(a) Acid Yellow 241

(b) Aci Yellow 194

(c) Acid Yellow 204

(d) Acid Orange 142

(e) Acid Red 357

(f) Acid Red 362

(g) Acid Violet 90

(h) Acid Violet 92

(i) Acid Blue 193

(j) Acid Brown 355

(k) Acid Brown 365

(l) Acid Brown 369

(m) Acid Green 104

(n) Acid Black 194

(o) Acid Black 172

Nowadays due to high labour cost and ecological structures, the European countries have stopped manufacturing metal-complex dyes and all are sourcing from China and India.

Ahmedabad is the hub for manufacturing acid dyes and reactive dyes. Some of the units import the cake from China in concentrated form and they do the mixing and pack it in drums with proper quality checking and providing technical service.

7.10   Stripping of metal-complex dyes

When necessity arises for stripping the dyes from the substrate due either to faulty dyeing or changes in style, this can be done by boiling of the dyed material with 5% ammonia (26%) for 20 min. followed by rinsing. This should be followed by another treatment in a fresh bath containing 3% formic acid (85%), raising the temperature to about 70°C, adding 4% Sulfoxite S Conc, followed by heating the bath to near the boil and treating under these conditions for about 45 minutes. Final rinse and neutralization with 3% ammonia (26%) at 50°C for 20 min complete the stripping process.

Acid dyes

8.1 Introduction

The first dye belonging to this class of dyes was made by Nicholson in 1862 by treating an insoluble dye, Violet Imperial (or Blue de Lyons), with sulphuric acid, when solubilising groups (sulphonic acid groups, SO_3H) were introduced in the dye, making it soluble in water. The resulting dye is known as Alkali Blue.

This reaction (sulphonation) is widely used to convert insoluble dyes into soluble ones. Thus, acidification of basic dyes led to the creation of acid dyes that are used mostly on wool and silk. For example, Magenta – a basic dye can be converted into Acid Magenta by sulphonation. They are also being more widely used for dyeing acetate, nylon, acrylics, modacrylics and spandex as well as blends of the above-mentioned fibers. These anionic dyes have also been found useful in printing chlorinated wool and silk. Approximately 80–85% of all acid dyes are sold to the US textile industry and are used for dyeing nylon, 10–15% for wool and the balance for those fibers mentioned above. Majority of acid dyes are sodium salt of aromatic sulphonic acid although some contains only carboxyl groups; the commercially available forms are usually their sodium salts, which exhibit good water solubility.

Acid dyes are commercially applied mainly on natural protein (wool and silk), synthetic polyamide (nylon) and to a small extent to acrylic and blends of these materials. Acid dyes are so-called because they are applied in presence of organic and inorganic acid in dyebath solutions. Generally, acid dyes do not have affinity for cellulosic fibers; however, there are some exceptions in this regard. Majority of acid dyes are sodium salts of aromatic sulphonic acids although some acid dyes contain carboxyl groups. The molecular weights of acid dyes range from 300 to 800. Acid dyes generally contain 1–4 sulphonic acid groups present in the dye structures.

Since sulphonic acid groups are strongly acidic, they exist in either solid or liquid form. Acid dyes are soluble in water and ionize into colored anions and colorless sodium ions. The anions are obviously large compared with cation and hence these dyes are also called anionic dyes. Acid dyes produce wide range of brilliant shades. Good light fastness of the dyed fabric can be obtained with selected dyes. These acid dyes are most suitable for loose wool fiber, wool tops, woolen hanks, knitted fabric, piece goods, carpet yarn and wool blend with cellulose yarn, synthetic yarn. These dyes possess brighter shades as well as good color fastness properties.

Acid dyes are commercially applied mainly on natural protein (wool and silk), synthetic polyamide (nylon) and to a small extent to acrylic and blends of these materials. Acid dyes are so-called because they are applied in presence of organic and inorganic acid in dyebath solutions. Generally, acid dyes do not have affinity for cellulosic fibers; however, there are some exceptions in this regard. Majority of acid dyes are sodium salts of aromatic sulphonic acids although some acid dyes contain carboxyl groups. The molecular weights of acid dyes range from 300 to 800. Acid dyes generally contain 1–4 sulphonic acid groups present in the dye structures. Since sulphonic acid groups are strongly acidic, they exist in either solid or liquid form. Acid dyes are soluble in water and ionize into colored anions and colorless sodium ions. The anions are obviously large compared with cation and hence these dyes are also called anionic dyes. Acid dyes produce wide range of brilliant shades. Good light fastness of the dyed fabric can be obtained with selected dyes.

Acid dyes are water-soluble anionic dyes that are applied to fibers such as silk, wool, nylon and modified acrylic fibers from neutral to acid dye baths. Attachment to the fiber is attributed, at least partly, to salt formation between anionic groups in the dyes and cationic groups in the fiber. Water-soluble acid dyes are not substantive to cellulosic fibers. Acid dyes are used both commercially and by the studio dyer to dye protein/animal fibers such as wool, silk, mohair, angora, alpaca and some nylons and synthetics. Acid dyes require the use of an acid such as vinegar, acetic or sulphuric acid to set the color. *An acid dye is a dye which is a salt of a sulfuric, carboxylic or phenolic organic acid.* The salts are often sodium or ammonium salts. Acid dyes are typically soluble in water and possesses affinity for amphoteric fibers while lacking direct dyes' affinity for cellulose fibers. When dyeing, ionic bonding with fiber cationic sites accounts for fixation of colored anions in the dyed material. Acids are added to dyeing baths to increase the number of protonated amino-groups in fibers.

Some acid dyes are used as food colorants. Most of them do not exhaust well on cellulosic fibers, but since they resemble direct dyes in chemical constitution, a reasonable number of them can dye cellulose quite well. A few such dyes are CI Acid Red 134, Blues 83 and 118, Blacks 7 and 48. However they are not commercially used for cellulose mainly because of poor wash fastness. Acid dyes are sold under various brand names, such as Sandolan E/N/P/Fast P/MF, (Clariant, formerly Sandoz), Acidol K/M, Palatine Fast (BASF), Supranol, Telon (Dystar, formerly Bayer), Polar (Ciba), Tulacid (Atul), etc.

Acid dyes sound scary to some novices, who imagine that the dyes themselves are caustic strong acids. In fact, the dyes are non-caustic, are in many cases non-toxic, and are named for the mild acid (such as vinegar) used in the dyeing process, and for the types of bonds they form to the fiber. Some of them are significantly more toxic than fiber reactive dyes, while others are even safe enough to eat, and are sold as food coloring.

Acid dyes are highly water soluble and have better light fastness than basic dyes. The textile acid dyes are effective for protein fibers such as silk, wool, nylon and modified acrylics. They contain sulphonic acid groups, which are usually present as sodium sulphonate salts. These increase solubility in water and give the dye molecules a negative charge. In an acidic solution, the $-NH_2$ functionalities of the fibers are protonated to give a positive charge: $-NH_{3+}$. This charge interacts with the negative dye charge, allowing the formation of ionic interactions. As well as this, Vander Waals bonds, dipolar bonds and hydrogen bonds are formed between dye and fiber. As a group, acid dyes can be divided into two sub-groups: acid leveling or acid milling. Acid dyes are thought to fix to fibers by hydrogen bonding, Vander Waals forces and ionic bonding. They are normally sold as the sodium salt therefore they are in solution anionic. Animal protein fibers and synthetic nylon fibers contain many cationic sites therefore there is an attraction of anionic dye molecule to a cationic site on the fiber. The strength (fastness) of this bond is related to the desire/chemistry of the dye to remain dissolved in water over fixation to the fiber.

Acid dyes are generally divided into three classes which depend on fastness requirements, level dyeing properties and economy. The classes overlap and generally depend on type of fiber to be colored as well as the process used.

Acid dyes affix to fibers by hydrogen bonding, Vander Waals forces and ionic bonding. They are normally sold as the sodium salt; therefore they are in solution anionic. Animal protein fibers and synthetic nylon fibers contain many cationic sites. Therefore, there is an attraction of anionic dye molecule to a cationic site on the fiber. The strength (fastness) of this bond is related to the tendency of the dye to remain dissolved in water over fixation to the fiber.

8.2   Chemical structure of acid dyes

According to their structure, acid dyes are mainly confined to three chemical classes as follows:

(1) Azo, (2) Anthraquinone, (3) Triphenylmethane

Although other classes namely nitro, indigoid, quinoline, azine, phthalocynanine, xanthane, carbolan dyes, etc., provide individually important classes of dye. Azo dyes represent the largest and most important group and are followed by anthraquinone and triarylmethane dyes. Of the other dye groups, very few products are of any commercial value. The proportion of acid dyes in each group is azo (65%), anthraquinone (15%), triphenylmethane (12%) and others (8 %). Acid dyes are obviously easy to apply and they allow the production of fast and often bright colors. Acid dyes are inexpensive and fairly fast to light, but they are not fast to washing and have only fair fastness to dry cleaning. They have low resistance to perspiration.

1. *Azo groups* – A very large number of acid dyes contain one or more azo groups. C.I. Acid Reds 1 and 142 are examples of mono and disazo dye, respectively. As the number of azo group within the molecule increases, the shades tend to become dull, darker and flatter. Hence, the number of trisazo acid dyes is limited. This class includes most of the yellows, oranges, scarlets and reds. The blue acid azo dyes are rather dull and not very important, except for the navy blues. Green and violet azo dyes are dull in shade. Brown azo dyes can be made by mixing. The azo class has a virtual monopoly for the black dyes.

2. *Anthraquinone groups* – They cover mostly blue hue range of acid dyes which give all bright blues. They are characterized by a high degree of wash fastness. Examples are C.I. Acid Blues 43, 45 and 127 and Red 80.

3. *Triphenylmethane groups* – This is the oldest class, the first member of this group; Nicholson Blue was first made in 1862. A few others are C.I. Acid Blues 1, 3 and 7 and Greens 5 and 50. This class provides brilliant blues, greens and violets. Among the other blues well known is C.I. Acid Blue 69. The greens include C.I. Acid Green 16 and the violets include C.I. Acid Violet 15. Dyes of this class have poor wash fastness; a few have moderate fastness.

In general, the chemical class of the dye does not determine its dyeing and fastness properties.

Acid dyes can be classified into four distinct classes as below:

(a) Leveling acid dyes

(b) Fast acid dyes

(c) Milling acid dyes

(d) Super-milling acid dyes

It is seen that the classification is mainly based on the requirement of acid added to the dyebath. More than half the acid, dyes are of the leveling type and of these about two-third are recommended only for dyeing from a sulphuric acid dyebath.

(a) *Leveling acid dyes (strong acid base dyeing)* – They are low molecular weight, contain sulphonic acid and require a highly acidic dyebath containing 3–5% sulphuric acid along with 10–20% (o.w.f.) Glauber's salt. They belong to molecularly split dyes and are highly soluble in water. As a class their light fastness are very good, but fastness to wet is not satisfactory. These dyes are used where superior leveling is of primary importance. Leveling dyes are primarily used for apparel, knit goods, carpets, upholstery, etc. During their applications, sulphuric acid acts as an exhausting agent, but if unevenness results, formic acid 80% (2–5%) may be used for more level dyeing. These dyes are dyed at pH 2.0–3.0. Dyeing is carried at boil for 45–60 minutes. These dyes have few limitations such as inferior color fastness to washing and low affinity to wool. These dyes seems poor color fastness to light especially in Turquoise Blue and pale shades.

Because dyeing of such dyes is carried out at pH 2.0–3.0 with mineral acid so prolong boiling during dyeing may create fiber degradation. They offer excellent leveling, migration and coverage of barre properties. Fastness to light is very good, while the wet fastness properties in heavy dark shades generally are only marginal. The latter can be improved with an after treatment of either tannic acid /tartar emetic or any other synthetic after-treating agent. The dyes of this group should be used when wet fastness properties are of no major concern and emphasis is put on good dyeing performance, such as coverage of barre. Typical representatives are C.I. Acid Yellow 49, C.I. Acid Red 337 and C.I. Acid Blue 40.

(b) *Fast acid dyes (moderate acid base dyeing)* – These dyes need strong acid assistance to dye but desirable pH is 4.0–4.5 instead of 2.0– 3.0. This pH is maintained either with help of acetic acid or ammonium sulphate. Mineral acid not required. For excellent leveling condition, formic acid 85% is recommended. Run the goods at boil for 45–60 minutes. Ekaline FI liquid may be added in dyebath as leveling agent. This class of dyes is highly brilliant in shade and tone. It possesses high tinctorial value. It yields excellent wet fastness properties and light fastness grading. For requirement of bright shades, this class of dyes is used successfully. Since these dyestuffs have excellent neutral affinity and good build up properties, they are suited especially for dyeing of medium to heavy dark shades. C.I. Acid Yellow 159 and C.I. Acid Red 299 are typical dyes of this class.

(c) *Milling acid dyes (weakly acid base dyeing)* – This class of dyes are used at pH 5.0–5.5. These dyes are well known for their general fastness properties to light, washing, water and wet fastness. These dyes are used commonly for dyeing of loose fiber, wool tops, woolen hank, woolen piece goods, wool–cotton, wool–polyester blends, etc. They are disulphonated dyes and provide dyeings with highest wet fastness properties. The leveling and migration properties of these products are much inferior to those of the monosulphonated dyes in (a) and (b) as above mentioned. Milling dyes generally lack the brightness and are used where good washing fastness is required. Milling dyes are suitable for use on wool that is to be milled into felt. The dyebath is set with 1–3% acetic acid (80%) and 10% Glauber's salt at 50°C, the goods are entered into the dyebath and the temperature is raised to boil within 45 minutes and dyed for 15–30 minutes depending on the depth of the shade. For complete exhaustion, 3% formic acid (85%) may be added at the end. C.I. Acid Yellow 79 is an example of this group.

(d) *Super-milling acid dyes (very weakly acid base dyeing)* – This is another behavior acid dyes used at pH 5.5–6.0. Dyes belong to this group need neutral pH so called 'Neutral Dyeing Acid dyes'. These dyes possess good affinity and excellent milling fastness so called 'Super Milling Acid dyes' commonly used for shading with metal-complex dyes. The super milling class of acid dyes is more hydrophobic due to long alkyl chains in the dye molecule giving very good wash and light fastness. These dyes have poor leveling properties and their applications need more care. It is dyed at boil for 45 minutes. 1–2% acetic acid (80%) may be added at the end to complete the exhaustion. The drawback of this method is that it requires long time of boiling. These dyes are highly fast for milling. No Glauber's salt is desired with these dyes as it works as retarding agent. Except ammonium sulphate or ammonium acetate, no further acid is needed to set the pH value. The major acid dyes manufacturers in India are Clariant and Colourtex. There are also lot of small suppliers and manufacturers dealing with acid dyes based in Mumbai, Indore and Ahmedabad. Acid dyes are having good scope in dyeing silk sarees and yarn and woolen hank, tops and loose fiber.

The chemistry of acid dyes is quite complex. Dyes are normally very large aromatic molecules consisting of many linked rings. Acid dyes usually have a sulfo or carboxyl group on the molecule making them soluble in water. Water is the medium in which dyeing takes place. These dyes are normally very complex in structure but have large aromatic molecules, having a sulphonyl or amino group which makes them soluble in water. Most of the acid dyes belong to following three main structural molecules:

1. Anthraquinon type
2. Azo dye type
3. Triphenylmethane type

Majority of commercially used acid dyes contain azo, anthraquinone and triarylmethane chromophore chemical composition. The proportion of acid dyes in each group is azo (65%), anthraquinone (15%), triphenylmethane (12%) and others (8%). Although there are other acid dyes, including azine, xanthane, nitro, indigoid, quinoline, phthalocyanine and carbalon dyes, these dyes are of limited commercial value.

The important chemical types are azo, anthraquinone and phthalocyanine, which cover the whole of the visible spectrum and thus give a complete color range. These dyes are soluble in water giving anionic species. They are usually applied at 100°C; whereas wool and other protein fibers degrade readily above this temperature, polyamide fibers (for example, the nylons) can be treated at 115°C without any harm coming to them.

The pH chosen for the solution in the dyebath depends on the individual properties of the dyes. The lower values are obtained by adding sulfuric acid and higher values by adding solutions of ethanoic acid and ammonium sulfate or ammonium ethanoate. Sodium sulfate may be added to control the diffusion of the dye anions in the fiber structure.

By the very nature of the dye structure, ionic bonds, hydrogen bonds and other intermolecular interactions will form between the dye and the fiber thus making the dyes fast. An example of a typical acid dye is C.I. Acid Red 73:

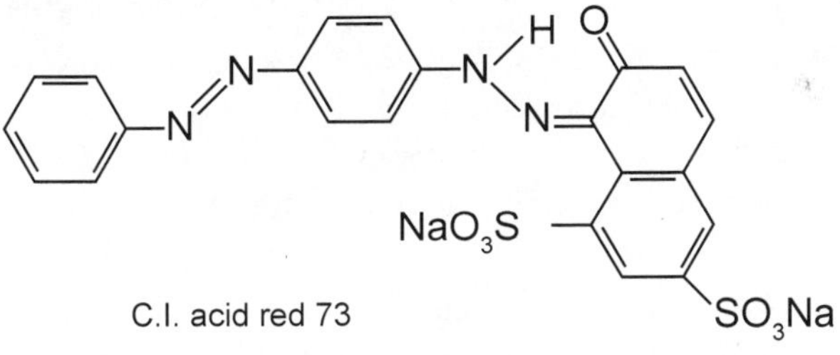

C.I. acid red 73

Figure 8.1

One of the azo groups in this tautomer is present as the ketohydrazone form.

8.2.1   Structures

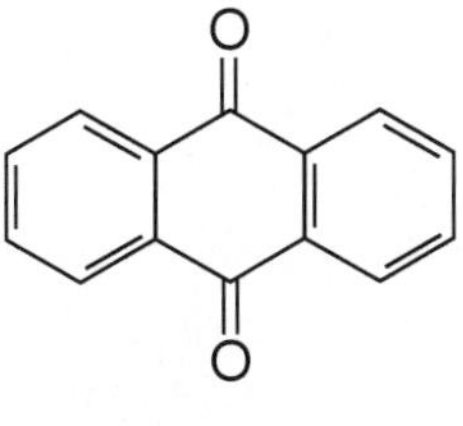

Figure 8.2

Anthraquinone derivatives generally form blue dyes.

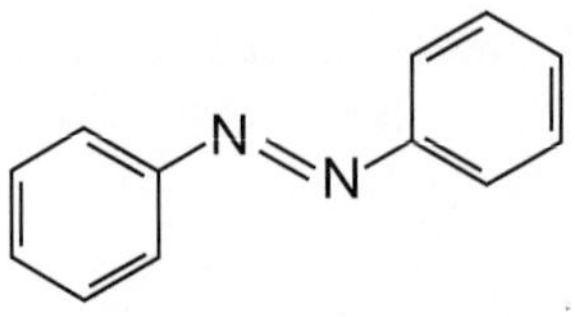

Figure 8.3

Azobenzene derivatives generally form red dyes.

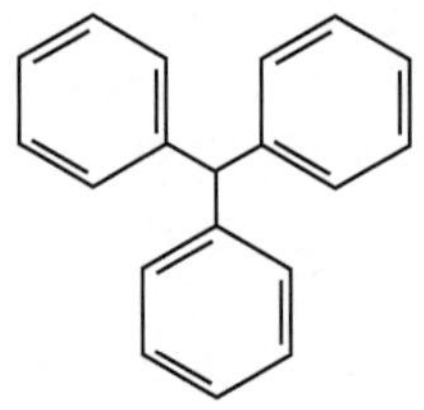

Figure 8.4

Triphenylmethane derivatives generally form yellow or green dyes.

The chemistry of acid dyes is quite complex. Dyes are normally very large aromatic molecules consisting of many linked rings. Acid dyes usually have a sulfo or carboxyl group on the molecule making them soluble in water. Water is the medium in which dyeing takes place. Most acid dyes are related in basic structure to the following:

Anthraquinone type: Many acid dyes are synthesized from chemical intermediates which form anthraquinone-like structures as their final state. Many blue dyes have this structure as their basic shape. The structure predominates in the leveling class of acid dye. Anthraquinone contains bright blue dyes with good light fastness. Some of the greens and violets are more brilliant than azo dyes. A typical chemical structure of anthraquinone class is Lissalamine Blue B (II).

Azo dyes: The structure of azo dyes is based on azobenzene, Ph–N=N– Ph. Although azo dyes are a separate class of dyestuff mainly used in the dyeing of cotton (cellulose) fibers, many acid dyes have a similar structure, and most are red in color. The first azo acid dye (I), Orange II, was formed by diazotizing sulphanilic acid and coupling with beta–naphthol.

Azo groups: A very large number of acid dyes contain one or more azo groups. C.I. Acid Reds 1 and 142 are examples of mono and disazo dye, respectively.

Triphenylmethane related: Acid dyes having structures related to triphenylmethane predominate in the milling class of dye. There are many yellow and green dyes commercially applied to fibers that are related to triphenylmethane. Triphenylmethane class provides brilliant blues, greens and violets.

Nitrite dyes are used for yellows and phthalocyanines for greenish blue.

8.2.2   Chemistry

Acid dyes are thought to attach to fibers by ionic bonds, hydrogen bonds, and Vander Waals forces. They are normally sold as the sodium salt; therefore they are in the form of anions in solution. Animal protein fibers and synthetic nylon fibers contain many cationic sites; therefore there is an attraction of the anionic dye molecule to a cationic site on the fiber. The strength (fastness) of this bond is related to the tendency of the dye to remain dissolved in water vis-a-vis its tendency to be fixed to the fiber.

The chemistry of acid dyes is quite complex. Dyes are normally very large aromatic molecules consisting of many linked rings. Acid dyes usually have a sulfonyl or amino group on the molecule making them soluble in water. Water is the medium in which dyeing takes place.

8.3   Manufacturing of acid dyes

1. Acid Orange II (Colour Index No. 15510)

    A solution of sulfanilic acid equivalent to 17.3 kg of 100% material in 200 litres water containing 6 kg of soda ash is boiled to drive off aniline. The solution is filtered, treated with 30 litres conc. hydrochloric acid and cooled to 20°C. The temperature is lowered to 10°C and diazotization is carried out by adding 7 kg of 100% sodium nitrite. The diazotization solution should give permanent test with congo red and starch iodide papers.

    While the diazotization is being carried out, 14.4 kg of beta-naphthol is dissolved in 200 litres of water containing 15 kg of 30% sodium hydroxide and 25 kg of soda ash.

    A clear solution should be obtained. This naphthol solution is cooled to 3°C and the suspension of diazosulfanilic acid is added in a thin stream. The temperature of the mixture should not rise above 8°C. After 1 hour, the reaction mixture is heated to boiling. The hot solution is treated with 100 kg of salt added portion wise. The precipitate which has now separated completely can easily be filtered at 50°C. The product is then pressed out and dried.

2. Acid Fast Red A (Colour Index No. 15620; Acid Red 88)

About 24.5 kg sodium naphthionate is dissolved in water and diazotized. The diazo-compound is filtered off and mixed with water to a thin cream. This is poured slowly into the beta-naphthol solution, which is prepared by heating the naphthol with a solution of 4.5 kg of caustic soda in 15 litres of water and pouring into 160 litres of cold water, and is stirred mechanically. The solution of sodium beta-naphtholate should not be above 15°C. When the whole of the diazo compound has been added, an alkaline reaction should be obtained with brilliant yellow paper. After an hour of stirring, the color is heated to 80°C and salt added until nearly the whole of the coloring matter is precipitated. This is then filtered, dried and ground.

8.4   Different types of acid dyes

The basic dyes are classified into several groups, based on the leveling properties, economy of the dyeing and fastness properties; however, generally these are classified into these three classes:

1. *Neutral acid dyes* – These are supra milling or fast acid dyes, having medium to good wet fastness properties, some of the dyes have poor light fastness in pale shades. Many of the dyes are used as self-shades only. These are applied to the fiber in a weakly acid or neutral pH.

2. *Weak acid dyes* – These dyes belongs to the milling class of dyes. These dyes have good fastness properties but light fastness is moderate to poor.

3. *Strong acid dyes* – These dyes are applied in a strongly acidic medium and also called leveling dyes; however, there wet fastness properties are a limitation. These dyes are very good to produce the combination shades.

8.4.1   Classification according to dyeing characteristics

Acid dyes are commonly classified according to their dyeing behavior, especially in relation to the dyeing pH, their migration ability during dyeing and their washing fastness. The molecular weight and the degree of sulphonation of the dye molecule determine these dyeing characteristics. The original classification of this type, based on their behavior in wool dyeing, is as follows:

(1) Level dyeing or equalizing acid dyes;

(2) Fast acid dyes;

(3) Milling acid dyes;

(4) Super-milling acid dyes.

Milling is the process in which a woolen material is treated, in weakly alkaline solution, with considerable mechanical action to promote felting. Dyes of good fastness to milling are essential to avoid color bleeding during the process.

8.4.2   Classes of acid dyes

Equalizing/leveling acid dyes: Highest level dyeing properties. Quite combinable in trichromatic shades. Relatively small molecule therefore high migration before fixation. Low wet fastness therefore normally not suited for apparel fabric.

Milling acid dyes: Medium to high wet fastness. Some milling dyes have poor light fastness in pale shades. Generally, not combinable. Used as self-shades only.

Metal complex acid dyes: More recent chemistry combined transition metals with dye precursors to produce metal complex acid dyes with the highest light fastness and wet fastness. These dyes are also very economical. They produce, however, duller shades.

8.4.3   Properties of acid dyes

The main properties of acid dyes are:
1. Since these are sold as a sodium salt, therefore, these form a large anion in the aqueous medium. 1. These dyes are anionic in nature.
2. These dyes are suitable for wool, silk, polyamide and modified acrylics. Acid dyes are used for dyeing protein fibers. The main protein fibers used for which acid dyes are used are wool, angora, cashmere and silk. Apart these, milk protein fibers like Silk Latte, Soya Protein, etc., can also be used.
3. These are applied from a strongly acidic to neutral pH bath.
4. These dyes have no affinity for cotton cellulose's, hence not suitable for cellulosics.
5. These dyes combine with the fiber by hydrogen bonds, Vander Waals forces or through ionic linkages.

8.4.4   Leveling acid dyes

Dyeing wool with leveling acid dye requires sulphuric or formic acid in the dyebath, along with Glauber's salt. Considerable amount of a strong acid is needed to achieve good exhaustion, typically 2–4% owf of sulphuric acid.

Because of the case of migration of leveling acid dyes during dyeing, the fastness to washing of their dyeing is only from poor to moderate. Their light fastness, however, ranges from fair to good. If the dye molecules do aggregate in solution at the maximum dyeing temperature, the aggregate are quite small, or there are enough individual molecules present in the solution for good penetration into the pores of the wool.

Wool contains about 820 mmol kg−1 of amino groups, some of which convert into ammonium ions in the presence of sulphuric acid, with a bound bisulphate anion. During dyeing, a dye anion displaces the bisulphate ion associated with an ammonium ion site. The wool is far from being saturated with dye anions. The added Glauber's salt acts as a retarding and leveling agent. It promotes leveling and reduces the dyebath exhaustion. As dyeing proceeds, more acid is then gradually added to decrease the bath pH. Leveling dyes give decreasing exhaustion on increasing the dyebath pH to values above 4, and with increasing temperature. These effects are consistent with a simple ion exchange process that is exothermic.

8.4.5   Fast acid dyes

These are usually monosulphonated acid dyes of somewhat higher molecular weight than typical leveling dyes. They dye wool by essentially the same dyeing method using acetic acid (1–3% owf) and Glauber's salt (5–10% owf). These dyes are used where level dyeing is necessary but when the washing and perspiration fastness of leveling acid dyes are inadequate.

8.4.6   Milling acid dyes

These anionic dyes have higher molecular weights and greater substantivity for wool than leveling or fast acid dye. They are medium to high wet fastness. Some milling dyes have poor light fastness in pale shades. Generally not combinable. Used as self-shades only.

8.4.7   Metal complex acid dyes

More recent chemistry combined transition metals with dye precursors to produce metal-complex acid dyes with the highest light fastness and wet fastness. These dyes are also very economical. They, however, produce duller shades.

8.5   Mechanism of dyeing with acid dyes

Dissolution of dyes in aqueous solvent produces a colored anion,

$$A\text{-}SO_3Na \xrightarrow{\quad H_2O \quad} A\text{-}SO_3^- + Na^+$$

Sodium salt   Colored

Of acid dye   Anion

Figure 8.5

    The protein and polyamide fibers produce cationic sites in water under acidic conditions; as the acidity of the solution is increased, more cationic sites are produced under these strongly acidic conditions. These cationic sites are thus available for the acid dye anions to combine with through hydrogen bonding, Vander Waals forces or ionic bonding. These linkages are strong enough to break, and thus dyeing produced is fast. Reaction between an acid dye and wool can be represented by following equation:

$$A\text{-}SO_3^{\ominus} + H_3N^{\oplus}\text{-}W\text{-}COOH \longrightarrow$$
$$A\text{-}SO_3^{\ominus}H_3N^{\oplus}\text{-}W\text{-}COOH$$

Acid dye on nylon at pH
$$A\text{-}SO_3^{\ominus}H_3\text{-}R\text{---}N^{\oplus}CO\text{---}R\text{-}COOH$$
$$^{\ominus}O_3S\text{-}A$$

Figure 8.6

8.5.1   On wool

Electrolyte in the acid dye bath acts as a retarding agent because of chloride ions attracted by the positive sites at the fiber and in the competition between. Addition of acid acts as an exhausting agent, because strongly acidic conditions make more cationic sites available and thus available dye anions got combined with these.

Dyeing temperature
The dyeing is generally carried out at boiling temperature for 30–60 min depending upon the depth of the shade and dyestuffs used.

Dyeing leveling agents
In the case of dyeing with acid dyes, mainly cationic agents such as ethoxylated fatty amines are used as leveling agents.

Heating rates
Heating rate is generally kept 1–3°C/min

Washing-off process
A typical dyeing cycle of nylon filament dyeing with acid dyes is shown in the below chart:

Wool dyeing method with acid dyes

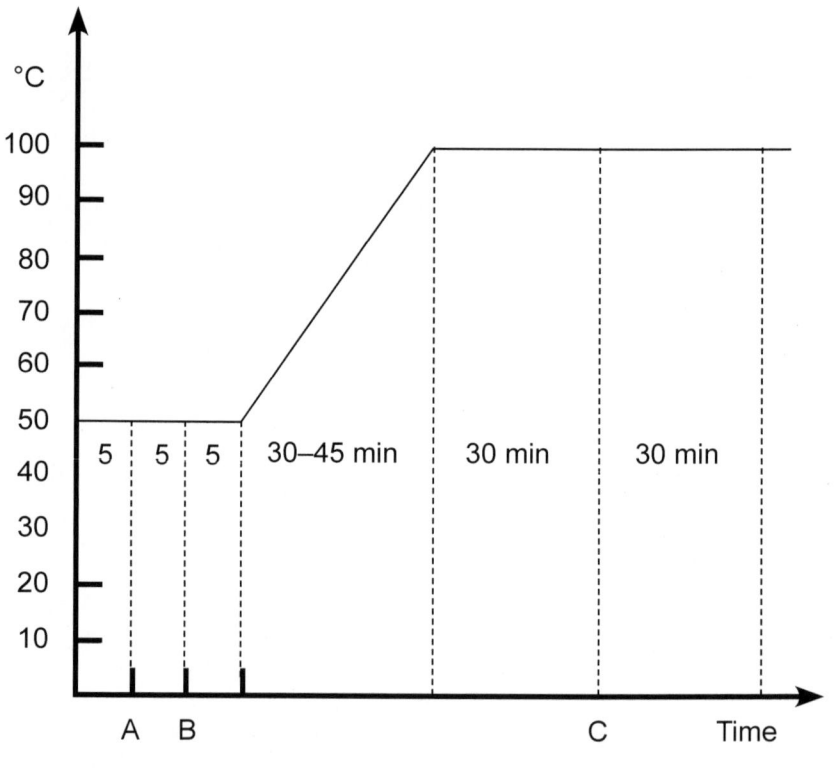

Figure 8.7

Method 1:
At (A), set bath at 50° with:
4% Sulphuric Acid (96%)
5% Glauber's Salt anhydrous
pH 2.5 to 3.5
At (B), add required amount to dyestuff

Method 2:

At (A), set bath at 50° with:

2% Formic Acid (85%)

5% Glauber's Salt anhydrous,

pH 3.5 to 4.5

At (B), add required amount of dye.

At (C), add 2% Sulphuric Acid (96%) or 2% Formic Acid (85%)

Thoroughly rinse after dyeing to remove loose color.

A dyeing cycle for nylon filament dyeing

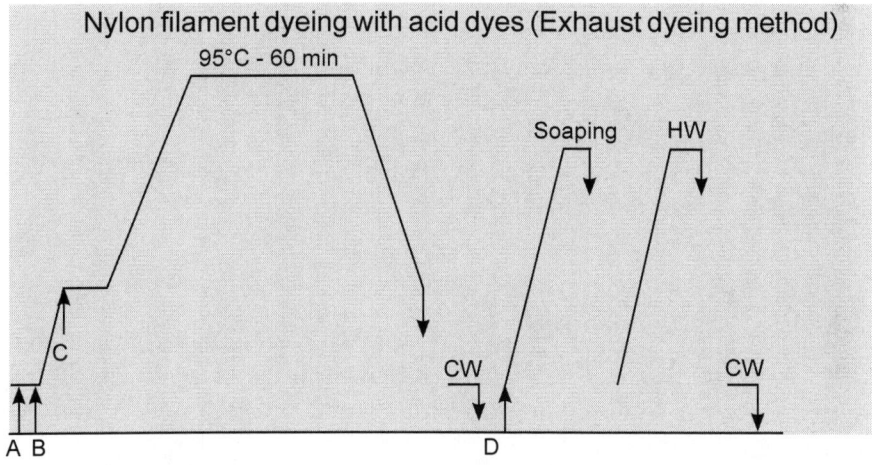

Figure 8.8

Fastness properties of acid dyes

The wet and light fastness properties of the acid dyes varies from poor to excellent, depending upon the molecular structure of the dyes.

The fastness properties as per the category are as follows:

Neutral acid dyes – Since these dyes have very good leveling and migration properties, and have a low affinity for the fiber, therefore the wet fastness properties of this class are generally poor.

    Weak acid dyes or half milling dyes – These dyes have a medium to good affinity for the fiber and are generally applied in a weakly acidic bath, shows medium to good wet fastness properties.

Strong acid dyes or super milling dyes – These dyes have poor exhaustion properties, therefore applied under very strong acidic condition, exhibit good fastness properties.

8.6   Identification of acid dyes

8.6.1   For wool and silk

A 100–300 mg portion of the dyed sample is placed in a 35 ml test tube, 5–10 ml of water and 0.5–1 ml of conc. ammonia are added and the mixture is boiled in order to bleed off a sufficient amount of dye for redyeing a piece of white cotton cloth.

The above part is same as acid dyes. However, if the sample in the direct dyes bled but left the test cotton white or only slightly stained, the colored extract is neutralized with 1 ml of 10% sulphuric acid solution and a few drops of acid are added in excess. A 20–30 mg of wool is added and the mixture is boiled for 1–2 min. The wool is rinsed and examined.

Redyeing of wool from an acid bath indicates the presence of acid dyes provided the presence of direct dyes is not shown.

8.7   Stripping of acid dyes

(a) With the molecularly dispersed acid dyes, on account of their good migrating properties, continued boiling in the same liquor is often effective. It must, however, be borne in mind that wool felts if it so boiled form too long.

(b) Boiling in fresh liquor with 20% of Glauber's salt will strip some of the color, and this can be exhausted again by cautious addition of acid or, preferably, ammonium salt.

(c) If treatment with Glauber's salt is not successful, more of the color can be stripped by boiling with 0.5% ammonia, or in liquor containing 2% pyridine.

(d) Another method is to boil the unevenly dyed material with 3–5% of a cationic, non-ionic mixture such as Tine gal W or Lyogen SMK. When sufficient color has been removed by any of these compounds, the wool is rinsed and redeye.

Basic dyes

9.1   Introduction

Only one basic dye, Beberine C.I. Natural Yellow 1+8, is known to occur in nature, but many of the earliest synthetic dyes including Perkin's Mauve, the first dye to be produced commercially from coal tar were basic. Perkin himself was largely responsible for devising the standard methods of application of these dyes to the fibres then available, namely wool, silk and cotton. Basic dyes like Magenta and Malachite green are also amongst the earliest synthetic dyes. In 1863 C. Martius first discovered cationic azo dye, Bismark Brown. Then in 1877 H. Caro patented Methylene Blue and O. Fischer synthesised Malachite Green. In 1924, BASF, Germany commercially introduced Cationic dyes. Then in 1938 IG Farben introduced Astrazon dyes. In 1975 Maxilan M dyes were made by Ciba and discontinued in 1990s. The basic dyes are so named because they are derived from organic bases. They are also called cationic dyes as they ionize in water-producing colored cations. Basic dyes are chiefly derived from azo, methane, triarylmethane, acridine, anthraquinone, azine, oxazine and xanthane derivatives. It is estimated that 50–100 different cationic dyes, drawing chiefly from azo, anthraquinone and TPM chromophore systems are now commercially available. Basic dyes are mainly applied the following type of fibres:

(a) Acidic types of acrylic fibres

(b) Mordanted cotton

(c) Cationic dyeable polyester fibre (CDP)

The first coal-tar dye was a so-called basic dye. It was developed to give many bright shades for silk and wool. The chemical agent that binds the dye to a fibre, which otherwise has little or no affinity for the dye, is known as a mordant. Basic dyes like Methyl violet, Auramine, Chrysodine, Methylene Blue, Magenta, Malachite Green and Rhodamine are used for dyeing Acrylic fibres in Hank form in Ludhiana. It is very cheap and cost effective. The Acrylic hank dyeing industry is very big in Ludhiana. They give good fastness and bright shades to acrylics for which they are principally used. But basic dyes when used on natural fibres are not fast to light, washing, perspiration or atmospheric gases, they tend to either bleed or crock. Till today in some old cotton process houses, they dye cotton fabrics using acid dyes but use basic dyes as an after-treatment for fabrics to improve the overall properties of the fabrics. It is also cheaper and cost effective. In some process house the cotton fabric is first treated with a suitable mordant then it is dyed with basic dyes. It is a long process and difficult to obtain level dyeings. This method can be used where fastness to light and washing is not that important. The most popular mordant used in textile units is Tannic acid (Tannin). It has to be noticed that the mordant does not affect the subsequent soaping process. Because Turkey-red oil and soap can also be used as mordants on cotton, they yield brighter shades than tannic acid but are not fast to soaping. The appeal of these basic dyes lays in their brilliant hues; some of them being fluorescent. Unfortunately their brilliance was not matched by their fastness. Because of their poor fastness properties, particularly to light, basic dyes were largely superseded following the development of other classes of dyes having superior fastness properties.

The importance of basic dyes has increased with the introduction of acrylic fibres in 1953. Although the original fibres of 100% polyacrylonitrile had poor dyeing properties, it was found that co-polymerisation with 5–10% of other monomers resulted in fibres of more open structure and could also be used to introduce anionic sites, thus, conferring ready dye-ability with basic dyes. Many existing basic dyes were found to be much faster to light on these fibres than on natural fibres. This led to the introduction of complete range specially designed for application to acrylic fibres. Moreover, the basic dyes showed much better fastness on these hydrophobic fibres as compared to those on natural wool and silk. The photo-fading of dyes is faster in the presence of water and oxygen, which cannot gain access in the hydrophobic polyacrylonitrile fibres. Though the light fastness of conventional basic dyes is better on acrylics, further improvements in dye structure lead to modified basic dyes, better known as Cationic dyes. These are not manufactured in India. They are imported by Jalan Company, Mumbai, from China, Korea, England, Thailand and Indonesia. Some of the ranges of cationic dyes manufactured for acrylic fibres by various manufacturers are Maxilon (Ciba), Astrazon (DyStar), Sevron, Yoracryl (YCL), Sandocryl B (Clariant).

In India, basic and cationic dyes are supplied by Mamta Dyes, Mumbai, Delhi and Ludhiana, Colourtex, Surat. Basic dyes are cheaper than Cationic dyes. The modacrylic fibres are which contain 35–85% polyacrylonitrile, those above 85% polyacrylonitrile being designated as acrylic fibres. Recent developments have been the introduction of anionic dye sites into nylon and polyester fibres for which few basic dyes are used. Such fibre is known CDP (Cationic dyeable polyester). It is mainly used in Surat for dyeing of fancy sarees, shirtings and dress material. They are very bright in shades as compared to normal polyester fibre dyed by disperse dyes. Basic dyes are also used for dyeing of paper and leather.

Basic dyes are used for dyeing of Bast fibres, Acetate fibres, Silk and Wool fibres. Basic dyes are used in printing of Acrylic fibres. A few basic dyes are soluble in cold or hot water, but most are hard to dissolve. This is due to small quantity of free base form being present. Usually most basic dyes are applied from pastes created by dispersing the basic dyestuff in acetic acid or alcohol.

Though basic dyes produce attractive, bright and highly intensive dyes, but their fastness to light and wet on mordanted cotton, protein fibres and acrylic fibres are very low. TPM dyes are substantive to acrylic fibres. Basic dyes applied on this fibre give bright dyeings with good fastness to light. This is because acrylic fibres are hydrophobic in nature and the dyesites in the fibre are not easily accessible to moisture and oxygen, which are responsible for fading process. Cationic dyes are available in:

(a) Powders

(b) Pearls

(c) Liquids

Apart from dyeing acrylics in the fibre, yarn or tow stage, dyeing can be produced at fibre production stage where basic dyes are used to wet spun acrylic tow in gel state (e.g. never dried state) prior to coagulation, drawing and steaming of the acrylic filaments (e.g. Courtaulds Neochrome process), which enables low batch weights of a specific colour to be produced economically on a continuous fibre production process. However, certain problems, such as uneven shades, loss in light fastness arise on gel dyeing.

Astrazon A dyes (DyStar) have the following cationic dyes for acrylic fibres.

1. Astrazon Yellow A-8GL

2. Astrazon Golden Yellow A-GL

3. Astrazon Red A-FBL

4. Astrazon Blue A-GRL

5. Astrazon Blue A-FBLS

6. Astrazon Blue A-BG

7. Astrazon Black A-FDL

They can be used for the following:

(a) Outer wear

(b) Hand knitting yarns

(c) Blankets

(d) Upholstery

(e) Decorative fabrics

The main advantages of Astrazon A dyes are the following:

1. A balanced range of colours from Bright yellow to jet black

2. High tinctorial strength

3. All necessary fastness requirements are achieved

For easier processing and to improve dependability of results, particularly in the exhaust dyeing of acrylic goods, the dyer needs to know a number of characteristics specific to the fibre and the cationic dye.

(a) *Fibre saturation value* (S) – The fibre saturation or S value indicates the maximum amount of dye which an acrylic fibre can take up. This is different for different acrylic fibres like IPCL (now Reliance), Pasupati, Indian Acrylic and imported acrylic fibre.

(b) *Saturation factor* (f) – The f value is a dye constant which is used to calculate the saturation concentration (Cs) in % of a given dye or combination of dyes obtainable on a given acrylic fibre. The formula used is as follows: Cs = S/f

(c) *Combination value/index* (K) – The K value indicates the compatibility of cationic dyes. Cationic dyes influence each other in their exhaustion so only dyes with the same or a similar combination index, i.e. a similar pick-up rate, should be combined in one recipe. The K values range from 1 to 5. In combination dyeings, dyes with a low K value exhaust onto the fibre first, while dyes with a higher value exhaust more slowly. The K values of dyes to be used in combination should not differ by more than 0.5–1.0. Retarders exhaust onto the acrylic fibre during the dyeing process and compete with the dyes, which are also cationic. This reduces the rate of the absorption of the dyes, thereby controlling uptake in the critical range and preventing unlevel dyeings.

All cationic dyes are stable over a given pH range. The pH for dyeing acrylic fibres is usually 4–4.5 and this should not be exceeded during the dyeing process. The dyebath pH is usually adjusted with acetic acid. Electrolytes like Glauber's salt, common salt have a retarding action and also promote migration. Cationic dyes start to exhaust at temperatures between 70°C and 85°C, depending on the type of fibre. The temperature raise should be very optimum. The dyes usually exhaust rapidly close to the boil. The cooling should be slow to achieve good dyeing. Acrylic tow and slubbing can be dyed efficiently and with good results by the pad-steam method. The tow dyeing machine is made from Serracant, Spain or Fleissner, Germany. Following is the procedure:

1. Padding with dyeliquor

2. Dye fixation by steaming

3. Washing off and softening (back washer)

4. Drying (perforated drum dryer)

Cationic dyes are applied by the continuous method. It must meet the following requirements:

(a)   Liquid or readily soluble powder form

(b)   Good build-up

(c)   Exhaustion on tone in combinations

(d)   High stability to steaming, including in combinations

Firstly the dye paste is made using cationic dyes, non-ionic, acid -resistant thickener (locust bean gum), non-ionic padding auxiliary, pH 4–4.5 with acetic or tartaric acid. Padding temperature: 30–50°C; steaming: 100–105°C; and drying: 110°C.

Basic dyes are also known as cationic dyes. This is a class of synthetic dyes that act as bases and when made soluble in water, they form a colored cationic salt, which can react with the anionic sites on the surface of the substrate. The basic dyes produce bright shades with high tinctorial values on textile materials. The appeal of the basic dyes lies in their brilliant hues, some of them being fluorescent. Unfortunately their brilliance is not matched by their fastness. The light and washing fastness properties of basic dyes are poor. Basic dyes are applied to wool and silk for brightness and tinctorial strength. The first synthetic dye mauve of Perkin (1856) belongs to this group of dyes. Basic dyes can be divided into following classification:

1. Diphenylmethane (e.g. Auramine)

2. Triphenylmethane (e.g. Malachite Green)

3. Azine

4. Thiazine (e.g. Methylene Blue)

5. Oxazine (e.g. Meldola's Blue)

6. Xanthene (e.g. Rhodamines)

Basic dye is a stain that is cationic (+ve charged) and so will react with material that is (−ve) negatively charged. This dye is usually synthetic that act as bases, and which are actually aniline dyes. Their color base is not water soluble but can be made so by converting the base into a salt. The basic dyes, while possessing great tinctorial strength and brightness, are not generally light-fast.

At the chemical level, basic dyes are typically cationic or positively charged. Basic dyes display cationic functional groups like $-NR_{3+}$ or $=NR_{2+}$. Since basic dye is a stain that is cationic or positively charged, it is the reason that it reacts well with material that is anionic or negatively charged.

Basic dyes are also known by name of cationic dyes and comprise class of synthetic dyes that works as base. Upon mixing in water, they assist in formation of colored cationic salt that react with anionic sites on surface of substrates. These basic dyes provide assistance in producing bright shades featuring superior tinctorial values on textile materials.

It is known as *Corantes Basicos* in Brazil.

It is known as *Colorantes Basicos* in Mexico, Colombia, Peru, Argentina, Paraguay, Uruguay, Chile, Guatemala, Honduras, Costa Rica, Nicaragua.

Basic dyes consist of amino groups, or alkylamino groups, as their auxochromes. Synthetic dye that was discovered by Perkin incidentally was a basic dye. Few examples of basic dyes are the following: methylene blue, crystal violet, basic fuchsin safranin, etc. An example of a basic dye that has amino groups as their auxochrome is pararosanilin or basic red 9 (according to the strict color index system of classification). Example of alkylamino groups is methylene blue or basic blue 9. Basic Blue 9 is a very popular dye that has vast use. A basic chemical structure of Basic Blue 9:

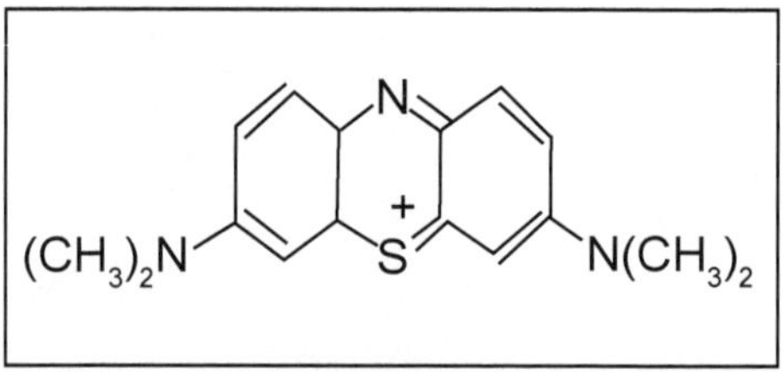

Figure 9.1

Basic dyes were amongst the earliest synthetic dyes. Indeed Mauveine is a basic dye. The chromophore is present as a cation and is used nowadays in dyeing acrylic fibres (usually a co-polymer with propenonitrile (acrylonitrile) and a small amount of a co-monomer which contain sulfonate, $-SO^{3-}$, and carboxylate, $-CO^{3-}$, groups). These are ion–ion interactions.

There are about 100 basic (cationic) dyes whose colours span reds, yellows and blues, with bright strong shades. Some are based on the azo and anthraquinone chromophore systems. Many are also based on arylcarbonium ions. Examples include C.I. Basic Green 4 (known as Malachite Green) and C.I. Basic Red 9.

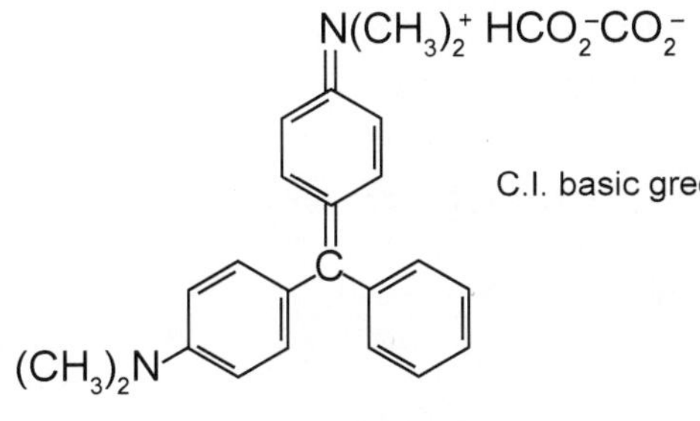
N(CH3)2+ HCO2-CO2-
C
(CH3)2N

C.I. basic green 4

Figure 9.2

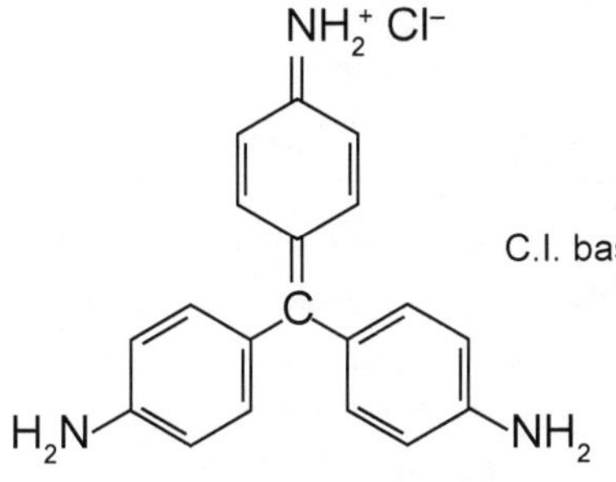
NH2+ Cl-
C
H2N
NH2

C.I. basic red 9

Figure 9.3

These are both triarylmethanes, a group of dyes, which with relatively small changes in structure produce a range of red, green and violet hues. Others, known as polymethine dyes (they contain one or more –CH= groups), are also used. They owe their colour to the presence of a conjugated system. An example of such a dye is C.I. Basic Yellow 28 which is a diazacyanine:

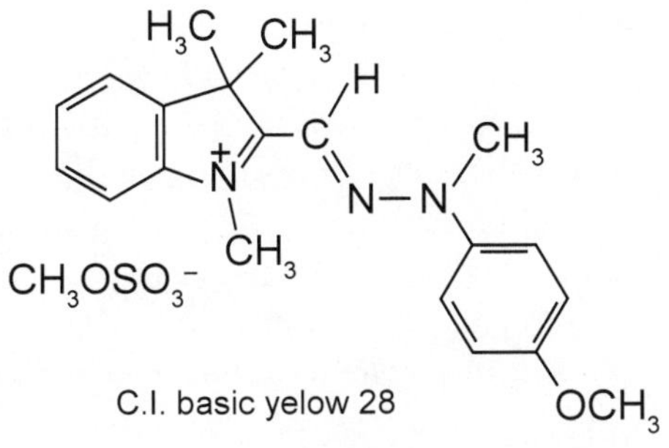
H3C
CH3
H
C
CH3
N
N–N
CH3
CH3OSO3-
OCH3

C.I. basic yelow 28

Figure 9.4

The dyes are often applied in a solution of an electrolyte, which controls the rate of diffusion in the fibre structure, at temperatures of about 370 K.

9.2   Manufacturing of basic dyes

1. *Auromine O* – It is the most important basic yellow and is highly valued because of the extraordinarily pure tints it produces. It is made by heating a mixture of 4,4'– dimethylamino diphenylmethane, sulphur and ammonium chloride and sodium chloride (acting as diluents) to 175°C in the presence of gaseous ammonia. The resulting Auromine base is converted into its hydrochloride which is Auromine O.

2. *Malachite Green* – It is prepared by condensation of two molecules of dimethylaniline with one of benzaldehyde. The leuco base, which is thus obtained, is then oxidized with oxidising agents of high electro chemical potential such as lead dioxide to give the dye salt with acid.

3. *Methylene Blue* – It is made by oxidizing a mixture of dimethylaniline and p-amino-dimethyl aniline (equimolar quantities) and sodium thiosulphate with sodium dichromate and hydrochloric acid in presence of zinc chloride (to get the final dye as zinc chloride double salt).

4. *New Magenta* – 1kg of anhydro formaldehyde, 5 kg of toluidine hydrochloride and 1 kg of o-toluidine are heated at 170°C for 2–3 hours with 1.2 kg of o-nitrotoluine and 100 grams of iron fillings, in an enameled vessel. The unreacted o-toludine and o-nitrotoluene are removed by distillation with steam, the residue is filtered hot, and the New Fuscsin is salted out.

9.3   Properties of basic dyes

Basic dyes are cationic-soluble salts of coloured bases. Basic dyes are applied to substrate with anionic character where electrostatic attractions are formed. Basic dyes are not used on cotton, as the structures are neither planar nor large enough for sufficient substantivity or affinity. *Basic dyes* are called cationic dyes because the chromophore in basic dye molecules contains a positive charge. The basic dyes react on the basic side of the isoelectric points. Basic dyes are salts, usually chlorides, in which the dyestuff is the basic or positive radical. Basic dyes are powerful colouring agents. It's applied to wool, silk, cotton and modified acrylic fibres. Usually acetic acid is added to the dyebath to help the take up of the dye onto the fibre. Basic dyes are also used in the coloration of paper.

Ionic nature: The ionic nature of these dyes is cationic.

Shade range: These dyes exhibit an unlimited shade range with high tinctorial strength, brightness and many colors are having fluorescent properties.

Solubility: The solubility of these dyes is very good in water, in the presence of glacial acetic acid.

Leveling properties: These dyes have a very high strike rate; therefore, leveling is poor.

Exhaustion: Cationic dyes exhaust at a variable rates, K values are used to define the exhaustion characteristics of the cationic dyes. K=1 means the fastest exhaustion, while K=5 means the slowest exhaustion. So while making the combination shades the dyes of similar K values must be used.

Affinity: These dyes show a very affinity towards wool, silk and cationic dye able acrylic, but have no affinity towards cellulosics. To dye cellulosics with basic dyes, the material must be treated with suitable mordanting agents.

- Work as cationic soluble salts of colored bases
- Can be applied to substrate with anionic character in areas where electrostatic attractions are formed
- Provides working as powerful coloring agents
- Can be applied to silk, wool, cotton and modified acrylic fibers
- Also used in coloration of paper
- Can offer varied shades with high tinctorial strength and brightness properties
- Having good solubility in water in presence of glacial acetic acid
- Relatively economical in usage
- Shows good brightness

Fastness property: The light fastness is poor to moderate, but wet fastness is good.

9.4   Dyeing of acrylic with basic dyes

The most common anionic group attached to acrylic polymers is the sulphonate group, $-SO_3-$, closely followed by the carboxylate group, $-CO_2-$. These are either introduced as a result of co-polymerisation, or as the residues of anionic polymerisation inhibitors. It is this anionic property which makes acrylics suitable for dyeing with cationic dyes, since there will be a strong ionic interaction between dye and polymer (in effect, the opposite of the acid dye-protein fibre interaction).

9.5   Advantages of basic dyes

- High Tinctorial strength
- Moderate substantivity
- Relatively economical
- Wide shade range
- Includes some of the most brilliant synthetic dyes
- Shows good brightness

9.6   Limitations of basic dyes

- Poor shade stability
- High acid content
- Coloured backwaters
- Very poor lightfastness
- Preferential dyeing

9.7   Modified basic dyes

These dyes, generally based on the chemistry of basic dyes, have longer molecular structures than traditional basic dyes, and thus have significantly improved properties. Though still cationic in nature, modified basic dyes exhibit improved fibre coverage and substantivity on many furnishes, making them ideal for dyeing applications. Lightfastness is also improved considerably over traditional basic dye.

9.8   Key advantages over conventional basic dyes

- Excellent substantivity
- Better Lightfastness
- Covers all fibres
- Clear backwaters

9.9   Limitations of basic dye

- High acid content
- Colored backwaters
- Poor shade stability

- Preferential dyeing
- Very poor light fastness

9.10   Modified basic dye

Modified basic dyes have similar chemistry, as of basic dyes. These only show a bit longer molecular structures than the typical conventional basic dyes, thus resulting in significantly improved properties.

9.10.1   Key advantages over conventional basic dyes

- Better Lightfastness
- Clear backwaters
- Covers all fibres
- Excellent substantivity

9.11   Application of the basic dyes

Basic dyes are extensively used for dyeing of jute, cut flowers, dried flower, coir, etc. For dyeing Acrylic Fibres, basic dyes are used widely. Modified basic dyes are used for dyeing of Acrylic Fibre, because these are perfect for this material. If the reason behind the success of Basic dyes is analysed, it would be seen that the positively charged cations of the Basic dyes gets attracted towards the negatively charged anions in the acrylic fibre. Acrylic polymers have anionic groups attached to it. They are most commonly the sulphonate group, $-SO_{3-}$, followed closed by the carboxylate group, $-CO_{2-}$. This reaction of the cation and anion results in salt linkages. Basic dye does not show absolutely any migration in acrylic fibers under normal dyeing conditions.

Basic dyes are also preferred to dye leather, because they can get combined easily with vegetable-tanned leather thus doing away with mordant. Basic dyes are also used in the coloring of papers.

Basic dyes for paper widely used in textile industries. Basic dyes offered by us are used mainly in the applications of acrylic fibers such as types 42 and 75 orlon and type 61 creslan. The dye is generally used to produce bright and deep shades with superior light and wash fastness. Its application is similar to that of direct dyes but requires different and more precise controls with auxiliaries and temperature.

9.12   Stripping of basic dyes

Stripping and leveling: A degree of leveling can be achieved by treatment with a carefully calculated amount of cationic retarder at the boil for 1–2 hours. For some stripping up to 10% owg of a suitable anionic retarder can be used also at the boil. For severe or chemical stripping, the goods should be treated with sodium hypochlorite acidified to pH ca. 6 with acetic acid for 30 minutes at the boil. Sodium nitrate (1 to 4 g/L) is required to passivate the stainless steel of the vessel.

10
Indigo dyes

10.1   Introduction

Historically speaking, natural Indigo dyes were first introduced in India, from where it eventually found its way to Europe. In fact, King George II is credited with the term "Navy Blue' as it was he who chose the indigo color for British naval uniforms. With such a rich past, it is no wonder that natural Indigo powder is widely used across the industry as a coloring agent. Indigo was used in India, which was also the earliest major center for its production and processing. The *I. tinctoria* species was domesticated in India. Indigo, used as a dye, made its way to the Greeks and the Romans, where it was valued as a luxury product.

Indigo is among the oldest dyes to be used for textile dyeing and printing. Many Asian countries, such as India, China, Japan, and Southeast Asian nations have used Indigo as a dye (particularly silk dye) for centuries. The dye was also known to ancient civilizations in Mesopotamia, Egypt, Britain, Mesoamerica, Peru, Iran, and Africa.

India is believed to be the oldest center of indigo dyeing in the old world. It was a primary supplier of indigo to Europe as early as the Greco-Roman era. The association of India with indigo is reflected in the Greek word for the dye, *indikón* (ινδικόν, Indian). The Romans latinized the term to *indicum*, which passed into Italian dialect and eventually into English as the word indigo.

Indigo dye is an important dyestuff with a distinctive blue color. The natural dye comes from several species of plant, but nearly all indigo produced today is synthetic. A variety of plants, including woad, have provided indigo throughout history, but most natural indigo is obtained from those in the genus *Indigofera*, which are native to the tropics. In temperate climate, indigo can also be obtained from woad (*Isatis tinctoria*) and dyer's knotweed (*Polygonum tinctorum*), although the Indigofera species yield more dye. The primary commercial indigo species in Asia was true indigo (*Indigofera tinctoria*, also known as *Indigofera sumatrana*). In Central and South America the two species *Indigofera suffruticosa* (Anil) and *Indigofera arrecta* (Natal indigo) were the most important.

Natural indigo was the only source of the dye until about 1900. Within a short time, however, synthetic indigo had almost completely superseded natural indigo, and today nearly all indigo produced is synthetic. Indigo is among the oldest dyes to be used for textile dyeing and printing. Many Asian countries, such as India, China, and Japan, have used indigo as a dye for centuries.

India is believed to be the oldest center of indigo dyeing in the old world. It was a primary supplier of indigo to Europe as early as the Greco-Roman era. The association of India with indigo is reflected in the Greek word for the dye, which was *indikon*. The Romans used the term *indicum*, which passed into Italian dialect and eventually into English as the word *indigo*.

The Romans used indigo as a pigment for painting and for medicinal and cosmetic purposes. It was a luxury item imported to the Mediterranean from India by Arab merchants. Indigo remained a rare commodity in Europe throughout the Middle Ages; woad, a dye derived from a related plant species, was used instead.

Figure 10.1 Woad

In the late fifteenth century, the Portuguese explorer Vasco da Gama discovered a sea route to India. This led to the establishment of direct trade with India, China, and Japan. Importers could now avoid the heavy duties imposed by Persian and Greek middlemen and the lengthy and dangerous land routes which had previously been used. Consequently, the importation and use of indigo in Europe rose significantly. Much European indigo from Asia arrived through ports in Portugal, the Netherlands, and England. Spain imported the dye from its colonies in South America. Many indigo plantations were established by European powers in tropical climates; it was a major crop in Jamaica and South Carolina. Indigo plantations also thrived in the Virgin Islands. However, France and Germany outlawed imported indigo in the 1500s to protect the local woad dye industry. Indigo was the foundation of centuries-old textile traditions throughout West Africa. The use of indigo here pre-dated synthetics. From the Tuareg nomads of the Sahara to Cameroon, clothes dyed with indigo signified wealth. Women dyed the cloth in most areas, with the Yoruba of Nigeria and the Manding of Mali particularly well known for their expertise. Among the Hausa male dyers working at communal dye pits were the basis of the wealth of the ancient city of Kano, and can still be seen plying their trade today at the same pits.

In Mesopotamia, a neo-Babylonian cuneiform tablet of the seventh century BC gives a recipe for the dyeing of wool, where lapis-colored wool (*uqnatu*) is produced by repeated immersion and airing of the cloth. Indigo was most probably imported from India. The Romans used indigo as a pigment for painting and for medicinal and cosmetic purposes. It was a luxury item imported to the Mediterranean from India by Arab merchants.

In Japan, indigo became especially important in the Edo period, when it was forbidden to use silk, so the Japanese began to import and plant cotton. It was difficult to dye the cotton fiber except with indigo. Even today indigo is very much appreciated as a color for the summer Kimono Yukata, as this traditional clothing recalls Nature and the blue sea.

In North America indigo was introduced into colonial South Carolina by Eliza Lucas Pinckney, where it became the colony's second-most important cash crop (after rice). When Benjamin Franklin sailed to France in November 1776 to enlist France's support for the American Revolutionary War, 35 barrels of indigo were on board the Reprisal, the sale of which would help fund the war effort. In colonial North America, three commercially important species are found: the native *I. caroliniana*, and the introduced *I. tinctoria* and *I. suffruticosa*.

Newton used "indigo" to describe one of the two new primary colors he added to the five he had originally named, in his revised account of the rainbow in *Lectiones Opticae* of 1675. Because of its high value as a trading commodity, indigo was often referred to as blue gold.

Indigo is a vat dye that is one of the oldest classes of dyes known. Until the late 19th century indigo was obtained from plants, the most significant *geing Indigofera tinctoria*. Experience, patience and sensitivity for subtle color changes are necessary for successful dyeing. Obtain the darkest blues through a properly reduced vat and repeated dyeing to build color depth. Synthetic Indigo requires a reducing agent, Thiox, and an alkali, Lye. When the fabric is removed from the yellow-green dye vat and begins to oxidize, the subtle changes from yellow-green through blue are captivating. In 1865 the German chemist Johann Friedrich Wilhelm Adolf von Baeyer began working with indigo.

His work culminated in the first synthesis of indigo in 1880 from o-nitrobenzaldehyde and acetone upon addition of dilute sodium hydroxide, barium hydroxide, or ammonia and the announcement of its chemical structure three years later. BASF developed a commercially feasible manufacturing process that was in use by 1897, and by 1913 natural indigo had been almost entirely replaced by synthetic indigo. In 2002, 17,000 tons of synthetic indigo were produced worldwide. The primary use for indigo is as a dye for cotton yarn, which is mainly for the production of denim cloth for blue jeans. On average, a pair of blue jean trousers requires 3–12 g of indigo. Small amounts are used for dyeing wool and silk.

Indigo carmine, or indigotine, is an indigo derivative which is also used as a colorant. About 20 million kg are produced annually, again mainly for blue jeans. It is also used as a food colorant, and is listed in the United States as FD&C Blue No. 2.

(1) *Natural Indigo* – A variety of plants have provided indigo throughout history, but most natural indigo was obtained from those in the genus Indigofera, which are native to the tropics. The primary commercial indigo species in Asia was true indigo (*Indigofera tinctoria*, also known as *I. sumatrana*). A common alternative used in the relatively colder subtropical locations such as Japan's Ryukyu Islands and Taiwan is *Strobilanthes cusia*. In Central and South America, the two species grown are *I. suffruticosa* (*añil*) and dyer's knotweed (*Polygonum tinctorum*), although the *Indigofera* species yield more dye.

Indigo dye is an important dyestuff with a unique shade of blue color. The natural dye comes from several species of plant. The dye gives a brilliant and eye-catching blue color to the fabric. This color partially penetrates the fabric but then also imparts surface blue color to the fabric. The word indigo is derived from the Latin Indicum and the Greek indikon meaning 'blue dye from India' or being more specific-'Indian substance'. Natural Indigo is the oldest known dye to mankind. When the synthetic substitute of the dye was not invented, all blue textiles were used to be dyed with indigo. These included the blue serge uniforms worn by the British police force and hospital staff, as well as military personnel and workman's clothes worn by millions of people, inspiring the term 'blue-collar worker'. The natural production of indigo is extracted from the leaves from a variety of plant species including indigo, woad, and polygonum. Only the leaves are used since they contain the greatest concentration of the dye molecules.

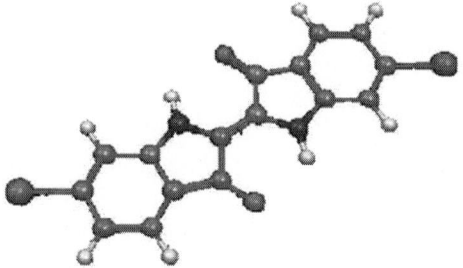

Figure 10.2   Structure

Indigo is a dark blue crystalline powder that melts at 390–392°C. It is soluble in chloroform, nitrobenzene, or concentrated sulfuric acid, but not soluble in water, alcohol and ether. The chemical structure/formula of indigo is $C_{16}H_{10}N_2O_2$. Indican is colorless and soluble in water which is a by-product of indigo. Indican is hydrolyzed to glucose and indoxyl. Indigo is converted to indoxyl by mild oxidation. When indigo is decomposed, simple compounds like aniline and picric acid are produced. Reduction of urea to indigo white is of great importance. The indigo white after it is applied to the fabric is reoxidized to indigo. History Indigo is one of the oldest dyes that were used for textile dyeing and printing. Asian countries like India, China and Japan had used indigo as a dye for centuries. The ancient civilizations in Mesopotamia, Egypt, Greece, Rome, Britain, Peru, Iran, and Africa had the knowledge about the dye. Since Neolithic times in Europe, this dye is being used. It was highly prized for its colour and light fastness. Until the end of the 19th century, the sole source was from plants, woad (*Isatis tinctoria*) and Dyer's Knotweed (*Polygonum tinctorum*) in temperate climates and *Indigofera* species in the tropics. Woad was widely grown in Europe, making some regions, especially Toulouse (France) and Erfurt (Germany), economically very prosperous. It was used to make a woad vat for dyeing with indigo from India after that period. It belongs to the legume family, and its three hundred plus species have been identified. In ancient times, indigo was considered a precious commodity because plant leaves contain only about small amount of the dye (about 2–4%). Indigo remained a rare commodity in Europe throughout the Middle Ages; woad, a dye derived from a related plant species, was used instead. Indigofera tinctoria was originally domesticated in India; it is mentioned in manuscripts dating from the 4th century BC about its origin in India. The blue serge uniforms worn by the British police force and hospital staff, as well as military personnel and workman's clothes worn by millions of people were dyed by indigo dye, the term blue-collar worker has been inspired from this aspect. It was a luxury item imported to the Mediterranean from India by Arab merchants.

The Portuguese explorer Vasco da Gama in late 15th century discovered a sea route to India. This led to the establishment of direct trade with India, the Spice Islands, China, and Japan. It resulted in avoidance of the heavy duties imposed by Persian, Levantine, and Greek middlemen on the importers and also the lengthy and dangerous land routes which had previously been used for transportation. As a result, the importation and use of indigo in Europe rose significantly. In the 19th century, the British obtained much indigo from India. When the synthetic substitute of indigo was invented the demand for natural indigo dropped, as a consequence for many indigo farmers the raising of indigo became loss-making business.

10.2   Cultivation of indigo dyes

Since synthetic indigo dye is being used, use of natural indigo dye has almost became extinct; but in recent years, the demand for natural dyes has been increasing in many countries again, because of health and pollution effects and a revival of interest in the relationship between dyes and culture. In present time, indigo is still cultivated for dye on a small-scale basis in India (particularly in the northern part of Karnataka) and in some parts of Africa and Central America. It is frequently grown as a secondary crop. Dissemination is usually by seeds which are sown at first in seedbeds or directly into the field. Germination takes about 4–5 days. When the plants are about 4–5 months old and starts flowering then the branches are harvested. At 3–4 months intervals three times a year the plants could be continued to be harvested. The total life span of the plants could range from time period 2–3 years.

10.3   Processing of indigo dyes

Figure 10.3

10.4   For natural indigo dye

The leaves of indigo go through a process of fermentation and then oxidized to yield the blue dye. Traditionally, fermentation was carried out by bacteria. The harvested plants, first of all, are packed into tanks and are covered with water. After a few hours of soaking the plants in water, the leaves become saturated and fermentation begins. A thick layer of bubbles and foam forms at the top of the tank. The process can be so vigorous that planks are placed on top of the tub to keep the plants in. This process can take up to a day and a half to complete, but the timing has to be perfect. The indigo makers will smell and taste the fluid to check whether the dye is ready or not. Even if an hour is taken extra it could ruin it. As soon as the liquid tastes sweet and is a dark blue colour, it is drained off into another vessel at a lower level, leaving the plants behind. The liquid now formed contains indoxyl.

- To stimulate oxidation of the indoxyl, the liquid is stirred continuously for many hours for it requires oxygen for oxidation. On alternative basis people will get into the vats to tread up and down to stir it up. After sometime the liquid turns a yellow-brown colour with floating dark blue patches. The solution is left to rest and the insoluble indigo settles to the bottom of the tank as a bluish sludge. To remove impurities and to stop the enzyme reaction which made the indigo the water is drained and filtered. The sludge is then dried to produce indigo 'cake' which is cut into cubes or made into balls.

- The Japanese used another method of extracting indigo from the polygonum plant. The plant is mixed with wheat husk powder, limestone powder, lye ash, and sake in this method of extracting indigo. The mixture is allowed to ferment for about one week to form a dye pigment which is called sukumo is formed from the fermented mixture.

- The indigo was used to be dissolved in stale urine in Europe as a pre-industrial process for dyeing in earlier times. The water-insoluble indigo is converted to soluble substance known as indigo white or leucoindigo by urine, which produces a yellow-green solution. After the indigo white oxidizes the fabric turns into blue color fabric.

- Another preindustrial method, used in Japan, was to dissolve the indigo in a heated tub in which a culture of thermophilic, anaerobic bacteria was maintained. Insoluble indigo could be converted into soluble indigo white by certain specific species of bacteria when they generate hydrogen as a metabolic product. Cloth dyed in such a vat was decorated with the techniques of shibori (tie-dye), kasuri, katazome, and tsutsugaki.

10.5   For synthetic production of indigo dye

- The synthesis reactions are conducted in large stainless steel or glass reaction containers. These containers hold jackets to permit steam or cold water to flow around the batch, as the reactions progress. On account of complication of these chemical processes, the dye is usually made in batch quantities. The Germans have invented methods for continuous process manufacturing.

- The first commercial method of producing indigo was based on Heumann's work. In this process, Nphenylglycine is treated with alkali to produce indoxyl, which could be then converted to indigotin by contact with air. But the amount of dye produced by this process is very less. More competent method, synthesis utilizes anthranilic acid. An alternate of this method (which is widely used) involves the reaction of aniline, formaldehyde, and hydrogen cyanide to form phenylglycinonitrile. This material is thereafter hydrolyzed to yield phenylglycine which is then converted to indigotin. Currently a method which uses sodamide with alkali to convert phenylglycine to indoxyl is used. Sodamide reacts with excess water, thus lowering the overall reaction temperature from almost 570°F (300°C) to 392°F (200°C). Efficient reaction process is achieved through it.

- After completion of the chemical reaction processes, the finished dye has to be washed to remove impurities and then dried. The dried powder then can be packed in drums or reconstituted with water to form a 20% solution and filled in containers.

Figure 10.4

10.5.1   Usage

- Indigo blue is the primary color of blue jeans in the west, its importance as result is increased in present time. In the United States, indigo is primarily used to dye cotton clothes and blue jeans. Over one billion pairs of jeans around the world are dyed blue with indigo. Deep navy blue colors on wool are produced by indigo dye.
- The Romans, in earlier times, used indigo as a pigment for painting and for medicinal and cosmetic purposes. It was a luxury item for Arabs which was imported to the Mediterranean from India by Arab merchants.
- Indigotinesulfonate is used as a dye in renal function testing and as a reagent for the detection of nitrates and chlorates. It is also used in the testing of milk.
- Indigo is also used as a food coloring under the name FD&C Blue No. 2.
- It was the original dye of the "Levi's" blue jeans, a trademark color for durability. It is the only natural blue dye that is permanent in nature.
- On many instances from Europe to the Arab world, historical use of blue to stain skin for war, religious and social rites were found.
- Indigo in combination with resist techniques creates decorative wooden items.
- In certain cultures indigo dye is also used for craft purpose such as on paper, leather and on silk.

10.5.2   The future

Scientists all over the world are carrying out research programmes on inventing a method of producing indigo dye which is less hazardous to health. Using biocatalysts in the dye reaction process might work out for producing environment-friendly dye. By making through a biological route, indigo dye would become one of the first high-volume chemicals by the biological route. Genencor International, of Rochester New York, is experimenting on a process to produce indigo using biotechnology. However, at this stage the technology is expensive and production costs might be prohibitive. The natural dye comes from several species of plant, but nearly all indigo produced today is synthetic in nature.

10.5.3   Extraction

The precursor to indigo is indican, a colorless, water-soluble derivative of the amino acid tryptophan. Indican readily hydrolyzes to release β-D-glucose and indoxyl. Oxidation by exposure to air converts indoxyl to indigo. Indican was obtained from the processing of the plant's leaves, which contain as much as 0.2–0.8% of this compound. The leaves were soaked in water and fermented to convert the glycoside indican present in the plant to the blue dye indigotin. The precipitate from the fermented leaf solution was mixed with a strong base such as lye, pressed into cakes, dried, and powdered. The powder was then mixed with various other substances to produce different shades of blue and purple.

10.5.4   Cultivation

Indigo was a major export crop that supported plantation slavery in colonial South Carolina in the 18th century. The demand for indigo in the 19th century is indicated by the fact that in 1897, 7,000 km^2 (2,700 sq m) were dedicated to the cultivation of indican-producing plants, mainly in India. By comparison, the country of Luxembourg is 2,586 km^2 (998 sq m).

Peasants in Bengal revolted against unfair treatment by the East India Company traders/planters in what became known as the Indigo Revolt in 1859, during the British Raj of India. In literature, the play *Nil Darpan* by Dinabandhu Mitra is based on the slavery and forced cultivation of indigo in India.

10.6   Synthetic indigo dye

Indigo dye is an organic compound with a distinctive blue color. Historically, indigo was a natural dye extracted from plants, and this process was important economically because blue dyes were once rare. Nearly all indigo dye produced today – several thousand tons each year – is synthetic. It is the blue of blue jeans.

In 1897, 19,000 tons of indigo were produced from plant sources. Largely due to advances in organic chemistry, production by natural sources dropped to 1,000 tons by 1914 and continued to contract. These advances can be traced to 1865 when the German chemist Adolf von Baeyer began working on the synthesis of indigo. He described his first synthesis of indigo in 1878 (from isatin) and a second synthesis in 1880 (from 2-nitrobenzaldehyde). The synthesis of indigo remained impractical, so the search for alternative starting materials at BASF and Hoechst continued. Johannes Pfleger and Karl Heumann eventually came up with industrial mass production synthesis. The synthesis of N-(2-carboxyphenyl) glycine from the easy to obtain aniline provided a new and economically attractive route. BASF developed a commercially feasible manufacturing process that was in use by 1897. In 2002, 17,000 tons of synthetic indigo were produced worldwide.

10.7   Pre-reduced indigo

This specially formulated indigo is already 60% reduced; therefore, it allows you to use soda ash instead of lye in the dye vat. Extremely easy to use, pre-reduced indigo makes setting up an indigo vat almost effortless. There is no need to paste up the indigo granules because they dissolve easily in water.

10.8   Natural indigo versus synthetic indigo

Although the chemical formula for natural and synthetic indigo is the same, synthetic indigo is almost pure indigotin. Natural indigo has a high proportion of impurities such as indirubins, which give beautiful colour variations. Like fine wines, the blue you get depends on where the indigo was grown and the weather at the time. Synthetic indigo, on the other hand, produces an even blue that never varies.

Natural indigo is a sustainable dye; after the pigment has been extracted the plant residue can be composted and used as a fertiliser and the water reused to irrigate crops. Natural indigo can often be traced to its country of origin, and even to the farm where it was produced. In buying it, you will be helping to give sustainable employment to rural population in developing countries. Synthetic indigo on the other hand is extracted from petrochemicals and its manufacture produces hazardous waste. By using natural indigo, you will be helping the environment and reducing the use of petrochemicals.

Natural indigo is the ideal blue dye to use on handmade textiles and on natural fibres; it may cost a little more than synthetic indigo, but the main cost of handmade items is time. Natural indigo is also essential for living history research, and for historic re-enactments. This may seem obvious, but if you want to use natural dyes, you need to use natural indigo rather than synthetic. Synthetic indigo is not a natural dye. Sometimes dyers may be unaware that they are in fact using synthetic indigo, as some shops don't always make it clear what type of indigo they are selling. If in doubt, check with your supplier.

10.9   Properties of indigo dyes

Indigo is a dark blue crystalline powder that sublimes at 390–392 °C (734–738 °F). It is insoluble in water, alcohol, or ether, but soluble in DMSO, chloroform, nitrobenzene, and concentrated sulfuric acid. The chemical formula of indigo is $C_{16}H_{10}N_2O_2$.

The molecule absorbs light in the orange part of the spectrum (λ_{max} = 613 nm). The compound owes its deep color to the conjugation of the double bonds, i.e. the double bonds within the molecule are adjacent and the molecule is planar. In indigo white, the conjugation is interrupted because the molecule is nonplanar.

Indigo dyestuff which is classified as vat dye is insoluble in water and has no affinity to the fibre. They have poor washing fastness which lets the color of denim fabric to change naturally. Indigo creates living colours on fabrics. Indigo dyestuff can never fully penetrate into the fibre, since its molecule is so big and it only adheres to the surface and remains at outer surface of the fibre. The inside stays white. It abrades or fades continually. This character of indigo lets denim fabric to have its final look with different types of washing and finishing applications. It enables denim fabric to response to finishing applications that gives a real life to the fabric. Indigo dye should be classified into two different chemical forms:

1. Natural form, insoluble in water (cannot dye the fibre)
2. Leuco form, soluble in water (can dye the fibre)

In natural form, indigo dyestuff has a color of blue but after reduced to leuco form, the color of the solution turns to yellow. What is reduction-oxidation?

In order that indigo is able to dye the fibre, it needs to be activated (leuco-form). In other words, indigo should be converted into soluble form and the affinity to fibre should be increased. Some chemical reactions are necessary for converting indigo to leuco form. These reactions are called "reduction". Reduction takes place in certain conditions with the presence of hydrosulfite ini alkaline medium. To keep the solution alkaline (basic), caustic (NaOH) is used.

After reducing and dyeing, dyed ropes have to be aerated so that the dye and fibre can be fixed together. This process is called "oxidation". However, the reduced leuco form of indigo has low affinity for the fiber. Therefore, more number of dips is required to achieve good indigo dyeing. Vatting is the chemical reduction process which is the origin of vat dyes. Penetration is the ability of dyestuff to diffuse or get into the fibre. Affinity is the attraction or force between dyestuff and fibre that causes them to combine.

10.10   Dyeing processes

Pad dry pad steam: This process can be performed with reactive indanthren and pigment dyestuff and has four main steps. At first step, dyestuff and auxiliary chemicals are fed into the dye pad and fabric picks up the dyestuff on itself. Second step is drying. After drying, fabric goes into the chemical pad at third step. Finally, dyestuff gets fixed on the fabric at the steamer. The amount of feeding and auxiliary chemicals might be changed according to the dyestuff used.

Pad steam: This process is performed with sulphur dye. "Pad-Steam" which is a part of PDPS dyeing method is used for this process. Fabric picks up chemicals and dyestuff from the same pad and goes to the steamer for fixation. Indigo dyeing is built on "continuous warp dying". Basically there are two main methods of indigo dyeing.

1. Classic method: Beaming, dyeing the warp yarns in rope form, rebeaming and sizing
2. Open-width method: Warp yarns are dyed and sized, respectively

Rope dyeing: This is the oldest way to dye warp yarns (ropes) and does not have any risk concerning "side to side" problem. Moreover, dyestuff absorption is almost the same since that all the ropes have the same tension during the process.

Loop dyeing: Warp yarns are dipped into the unique pad many times.

Slasher dyeing: Warp yarns are dyed as open-width form and dyeing, drying, sizing processes are performed in the same machine continuously.

10.11   Advantages of rope dyeing against slasher dyeing

•   Large quantities can be dyed continuously.
•   In rope dyeing, ropes are dipped into the dye pads with identical tension and angle; therefore, there is not any risk of "side to side" problem.
•   Dyeing machine does not have to stop while feeding new dyeing parties which means energy saving.
•   Yarn wastage is not that much.

10.12   Indigo dyeing process flow

1. *Pre-processes (pre-process pads)*: According to the desired final look and dyeing properties, pre-wetting, bottom dyeing or washing can be achieved as a pre-process. Pre-Wetting helps to increase the affinity of the warp yarn to the dyestuff. Bottom dyeing is needed to get different casts using reactive, indanthren or sulphur dyestuffs. Washing After bottom dyeing process, ropes should be washed in order to remove excess dyestuff unfixed from the warp

(rope). This is crucial for optimum crocking values. Pre-wetted ropes are ready to be dipped into dye pads. Dipping and aerating (oxidizing) are repeated until required cast is achieved.

2. *Dyeing process (dye pads):* Affinity of indigo dyestuff is still not sufficient after reduction. Thus, indigo dyeing process is based on repeated dipping and aeration. Dyeing machines are designed taking this point into consideration. That's why they have more than one dye pad. Basically, ropes continuously take the dyestuff from the pads and are fixed thanks to air reaction.

Indigo dyeing has three important parameters:

- Indigo
- Hydrosulfite (used as reduction agent)
- Caustic (to keep the pH of the bath as alkaline)

These chemicals must be fed at sufficient ratio consistently. Dye pads are supported by an internal circulation system. The amount of chemicals to be fed must be constant during the process in order to avoid indigo-hydro concentration and pH value differences. This circulation system eliminates possible shade differences on fabric.

10.13   What's ring dyeing?

As already known, indigo dyed ropes (warp yarns) have an ecru core and its outer layer is dyed. This is called "ring-dyeing". The ring-dyeing property of a rope is determined by pH value and hydro concentration. The penetration of dyestuff into fibre in a dye bath having 11.5 pH is not good. When pH value shifts towards 13.0, in other words increases, penetration gets better; dyestuff can better penetrate into fibre core, that's to say, ring dyeing efficiency drops. This drastic drop affects fabric-washing properties. Efficiently ring-dyed fabrics would respond washing (finishing) better and faster. These parameters have to be adjusted according to end-product properties. The rope turns yellow-green colour when first dipped into indigo bath. It turns to magic indigo blue as soon as it reacts with air. This legendary colour change is really worthwhile.

Washing and drying of ropes (washing pads and drying cylinders) – After iterative numbers of dipping, ropes follow washing pads in order to remove unfixed indigo dyestuff. Besides, this step has a neutralization effect, as pH value needs to be lowered because of high pH environment during dyeing process. This step is completed reaching sufficient humidity with the help of steamed drying cylinders. Sufficient humidity ratio is important for the efficiency of re-beaming of ropes after dyeing.

Main parameters of indigo dyeing:

- *Reduced indigo concentration*: Effective on colour depth and darkness.
- *Hydro concentration*: It is the chemical that helps reduction of indigo dyestuff. Since reduced indigo has strong decomposition tendency, there must be excess hydro in dye bath. The preservation of reduced form is accomplished by excess hydro. Therefore, the control of hydro concentration in dye bath solution has great importance. The amount of hydro is affects penetration of indigo dye into fiber.
- *pH* (*the alkality of solution*): The typical feature of indigo dyestuff is that pH should be higher than 11.5. The best dyeing is achieved between 11.5–12.5 pH. The penetration increases when pH is increased, darker and consistent shades can be obtained. Therefore, the response to stone-washing gets more difficult.
- *Number of dye pads*: This is effective on shade depth.
- *The speed of machine is effective on following parameters*: 1. Dipping time; 2. Penetration; 3. Oxidation (airing) time.
- *Pressure* (*Nip*) *rollers affect following parameters*: 1. Amount of dyestuff picked up; 2. Penetration.

10.14   Manufacturing of indigo dyes

Given its economic importance, indigo has been prepared by many methods. The Baeyer-Drewson indigo synthesis dates back to 1882. It involves an aldol condensation of o-nitrobenzaldehyde with acetone, followed by cyclization and oxidative dimerization to indigo. This route is highly useful for obtaining indigo and many of its derivatives on the laboratory scale, but was impractical for industrial-scale synthesis. Johannes Pfleger and Karl Heumann eventually came up with industrial mass production synthesis. The first commercially practical route is credited to Pfleger in 1901. In this process, N-phenylglycine is treated with a molten mixture of sodium hydroxide, potassium hydroxide and sodamide. This highly sensitive melt produces indoxyl, which is subsequently oxidized in air to form indigo. Variations of this method are still in use today. An alternative and also viable route to indigo is credited to Heumann in 1897. It involves heating N-(2-carboxyphenyl) glycine to 200°C (392°F) in an inert atmosphere with sodium hydroxide. The process is easier than the Pfleger method, but the precursors are more expensive. Indoxyl-2-carboxylic acid is generated. This material readily decarboxylates to give indoxyl, which oxidizes in air to form indigo. The preparation of indigo dye is practiced in college laboratory classes according to the original Baeyer-Drewsen route.

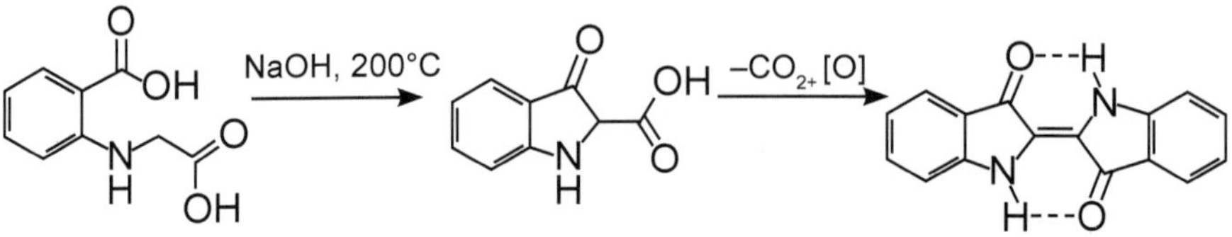

Figure 10.5 Heumann's original synthesis of indigo

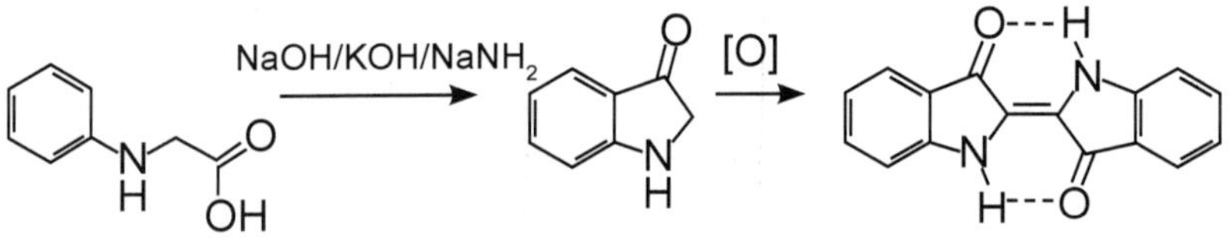

Figure 10.6 Pfleger's synthesis of indigo

10.15   Manufacturing process of Denim

The term "Denim" has originated from the city of Nimes in France where "serge de Nimes" was manufactured. Denim is made from a vat dye, the Indigo dye, which is applied to cotton fabric in loosely held form in layers. As far as manufacturing process of *denim* is concerned, it is similar to that of grey fabric up to the process of weaving with the only difference that in case of Denim fabric, it is dyed at the stage of sizing where as in case of grey fabric, the decision regarding dyeing stage depends upon the finished product.

Figure 10.7

10.15.1   Spinning

The initial processes of denim manufacturing consist of the regular activities of opening and blending of cotton fibers. Carding is done to remove any foreign matter and the short fibers so that cotton takes the form of a web which is then converted into a rope-like form, the sliver. Then drawing process produces a single, uniform sliver from a number of carded slivers. Yarn is then spun through open-end spinning or ring spinning. Roving is also carried on, if the spinning has to be done through ring spinning. Generally, denim fabric are 3/1 warp-faced twill fabric made from a yarn-dyed warp and an undyed weft yarn. Normally dyed and grey ring or open-end yarns are used in warp and weft, respectively. Traditionally speaking, the warp yarn is indigo dyed.

Figure 10.8

10.15.2   Warp preparation: Dyeing and sizing processes

Warp yarns are indigo dyed and sized with the help of two methods.

(i)   Threads from several back beams are combined to form a warp sheet and dyed and sized on the same machine.

(ii) Threads, about 350–400 in number, are formed into ropes. 12–14 ropes run adjacent to each other through the continuous dyeing unit. After dyeing, the ropes are dried on drying cylinders and then collected in a can. After that, a worker's beam is prepared. Sizing is then done in the conventional manner.

Figure 10.9

There are various dyeing and sizing processes, which can be classified into four categories.

- Continuous indigo-rope dyeing and sizing
- Continuous indigo dyeing and sizing
- Indigo-back beam dyeing and sizing
- Continuous dyeing and sizing

10.15.3   Continuous indigo-rope dyeing and sizing

The yarn coming out from the ring frames is wound into cheeses or cones and then placed on the ball warper on which 350–400 threads are formed into a rope and are cross wound to a ball in accordance with the length of warp beam. During this process, lease bands are inserted at particular intervals as they are required for further processing on long chain beamer. Based on the size of the rope dyeing plant, 12–24 ropes, at a time, are dyed, oxidized, dried and placed in large containers. These ropes are then opened on the long chain beamer through tension roll and expansion comb and wound on to a back beam. Back beams are then sized and the sized warp is then woven. This system is commonly used in the United States.

Figure 10.10

10.15.4   Continuous indigo dyeing and sizing

In this process, back beams are processed on the dyeing/sizing machine instead of ropes. The warp is dyed, oxidized, dried and sized at a one go. Although this process is less cumbersome, the risk of individual thread breakage is greater than dyeing in rope form. This method is commonly used in the European countries.

10.15.5   Indigo-back beam dyeing and sizing

Dyeing and sizing is done in two stages in this method. In the first stage, back beams are dyed, oxidized, dried and wound on a batch roll. The batch roll is then sized, dried and wound on a weaver's beam.

10.15.6   Continuous dyeing and sizing

Although glass is hard and rigid, yet it can be transformed into fine, translucent and flexible glass fiber, commonly known as fiberglass. It is very glossy in appearance and feels like silk. There are two methods for glass fiber manufacturing: Continuous filament process and staple fiber process. Apart from being glossy and flexible, glass fiber is also heat resistant. Due to its many qualities, this fiber is widely used for home furnishings, apparels and many other industrial purposes. It's really very interesting to know about the whole process of glass fiber manufacturing.

10.15.7   Weaving

The weaving process interlaces the warp, which are the length-wise indigo dyed yarn and the filling, which are the natural-colored cross-wise yarn. The warp thread is in the form of sheet. The weft thread is inserted between two layers of warp sheets by means of a suitable carrier, such as shuttle, projectile, rapier, air current, water current, etc. The selection of carrier depends upon the type of *weaving machinery* used. The two different technologies available for weaving machines are the following: Conventional shuttle weaving system which is done by ordinary looms or automatic looms; and the shuttle less weaving system which is done by airjet, waterjet, rapier, or a projectile weaving machine. The conventional shuttle loom results in lesser production due to slow speed and excessive wear and tear of machinery. As such, now denim is generally woven through shuttle-less weaving system namely, airjet looms, rapier looms or projectile looms.

Figure 10.11

10.15.8 Finishing

The final woven fabric, wound on a cloth roll, is taken out from weaving machines at particular intervals and checked on inspection machines so that any possible weaving fault can be detected. In this quality control exercise, wherever any fault is seen, corrective measures are taken then and there only. The woven denim fabrics then goes through various finishing processes, such as brushing, singeing, washing, impregnation for dressing and drying. Brushing and singeing eliminate impurities and help to even the surface of denim fabric. Dressing regulates the hand and rigidity of the fabric while compressive shrinking manages its dimensional stability. The standard width of denim fabrics is then sent for making up. In this process, the fabric is cut into the desired width according to the size required. The made-up denim fabric is then thoroughly checked for defects such as weaving defects, uneven dyeing, bleaching and dyeing defects, oil stains, or patches. After inspection, the final product is categorized quality-wise. The fault-less fabrics are sent to the packaging department while the defective ones are sent for further corrections.

Figure 10.12

Pigment dyes

11.1   Introduction

Naturally occurring pigments such as ochres and iron oxides have been used as colorants since prehistoric times. Archaeologists have uncovered evidence that early humans used paint for aesthetic purposes such as body decoration. Pigments and paint-grinding equipment believed to be between 350,000 and 400,000 years old have been reported in a cave at Twin Rivers, near Lusaka, Zambia.

Before the Industrial Revolution, the range of colors available for art and decorative uses was technically limited. Most of the pigments in use were earth and mineral pigments, or pigments of biological origin. Pigments from unusual sources such as botanical materials, animal waste, insects, and mollusks were harvested and traded over long distances. Some colors were costly or impossible to mix with the range of pigments that were available. Blue and purple came to be associated with royalty because of their expense.

Biological pigments were often difficult to acquire, and the details of their production were kept secret by the manufacturers. Tyrian Purple is a pigment made from the mucus of one of several species of Murex snail. Production of Tyrian Purple for use as a fabric dye began as early as 1200 BCE by the Phoenicians, and was continued by the Greeks and Romans until 1453 CE, with the fall of Constantinople. The pigment was expensive and complex to produce, and items colored with it became associated with power and wealth. Greek historian Theopompus, writing in the 4th century BCE, reported that "purple for dyes fetched its weight in silver at Colophon [in Asia Minor]."

Mineral pigments were also traded over long distances. The only way to achieve a deep rich blue was by using a semi-precious stone, lapis lazuli, to produce a pigment known as ultramarine, and the best sources of lapis were remote. Flemish painter Jan van Eyck, working in the 15th century, did not ordinarily include blue in his paintings. To have one's portrait commissioned and painted with ultramarine blue was considered a great luxury. If a patron wanted blue, they were obliged to pay extra. When Van Eyck used lapis, he never blended it with other colors. Instead he applied it in pure form, almost as a decorative glaze. The prohibitive price of lapis lazuli forced artists to seek less expensive replacement pigments, both mineral (azurite, smalt) and biological (indigo).

The son of a master dyer, Tintoretto used Carmine Red Lake pigment, derived from the cochineal insect, to achieve dramatic color effects.

Spain's conquest of a New World empire in the 16th century introduced new pigments and colors to peoples on both sides of the Atlantic. Carmine, a dye and pigment derived from a parasitic insect found in Central and South America, attained great status and value in Europe. Produced from harvested, dried, and crushed cochineal insects, carmine could be, and still is, used in fabric dye, food dye, body paint, or in its solid lake form, almost any kind of paint or cosmetic.

Natives of Peru had been producing cochineal dyes for textiles since at least 700 CE, but Europeans had never seen the color before. When the Spanish invaded the Aztec empire in what is now Mexico, they were quick to exploit the color for new trade opportunities. Carmine became the region's second most valuable export next to silver. Pigments produced from the cochineal insect gave the Catholic cardinals their vibrant robes and the English "Redcoats" their distinctive uniforms. The true source of the pigment, an insect, was kept secret until the 18th century, when biologists discovered the source.

While Carmine was popular in Europe, blue remained an exclusive color, associated with wealth and status. The 17th-century Dutch master Johannes Vermeer often made lavish use of lapis lazuli, along with Carmine and Indian yellow, in his vibrant paintings.

11.2   Development of synthetic pigments

The earliest known pigments were natural minerals. Natural iron oxides give a range of colors and are found in many Paleolithic and Neolithic cave paintings. Two examples include Red Ochre, anhydrous Fe_2O_3, and the hydrated Yellow Ochre ($Fe_2O_3.H_2O$). Charcoal, or carbon black, has also been used as a black pigment since prehistoric times.

Two of the first synthetic pigments were white lead (basic lead carbonate, $(PbCO_3)_2 Pb(OH)_2$) and blue frit (Egyptian Blue). White lead is made by combining lead with vinegar (acetic acid, CH_3COOH) in the presence of CO2. Blue frit is calcium copper silicate and was made from glass colored with a copper ore, such as malachite. These pigments were used as early as the second millennium BCE.

The Industrial and Scientific Revolutions brought a huge expansion in the range of synthetic pigments, pigments that are manufactured or refined from naturally occurring materials, available both for manufacturing and artistic expression. Because of the expense of Lapis Lazuli, much effort went into finding a less costly blue pigment.

Prussian Blue was the first modern synthetic pigment, discovered by accident in 1704. By the early 19th century, synthetic and metallic blue pigments had been added to the range of blues, including French ultramarine, a synthetic form of lapis lazuli, and the various forms of Cobalt and Cerulean Blue. In the early 20th century, organic chemistry added Phthalo Blue, a synthetic, organometallic pigment with overwhelming tinting power.

Working in the late 19th century, Cézanne had a palette of colors that earlier generations of artists could only have dreamed of. Discoveries in color science created new industries and drove changes in fashion and taste. The discovery in 1856 of mauveine, the first aniline dye, was a forerunner for the development of hundreds of synthetic dyes and pigments like azo and diazo compounds which are the source of a wide spectrum of colors. Mauveine was discovered by an 18-year-old chemist named William Henry Perkin, who went on to exploit his discovery in industry and become wealthy. His success attracted a generation of followers, as young scientists went into organic chemistry to pursue riches. Within a few years, chemists had synthesized a substitute for madder in the production of Alizarin Crimson. By the closing decades of the 19th century, textiles, paints, and other commodities in colors such as red, crimson, blue, and purple had become affordable.

Development of chemical pigments and dyes helped bring new industrial prosperity to Germany and other countries in northern Europe, but it brought dissolution and decline elsewhere. In Spain's former New World empire, the production of cochineal colors employed thousands of low-paid workers. The Spanish monopoly on cochineal production had been worth a fortune until the early 19th century, when the Mexican War of Independence and other market changes disrupted production. Organic chemistry delivered the final blow for the cochineal color industry. When chemists created inexpensive substitutes for carmine, an industry and a way of life went into steep decline.

11.3   New sources for historic pigments

Vermeer was lavish in his choice of expensive pigments, including Indian Yellow, lapis lazuli, and Carmine, as shown in this vibrant painting. Before the Industrial Revolution, many pigments were known by the location where they were produced. Pigments based on minerals and clays often bore the name of the city or region where they were mined. Raw Sienna and Burnt Sienna came from Siena, Italy, while Raw Umber and Burnt Umber came from Umbria. These pigments were among the easiest to synthesize, and chemists created modern colors based on the originals that were more consistent than colors mined from the original ore bodies. But the place names remained.

Historically and culturally, many famous natural pigments have been replaced with synthetic pigments, while retaining historic names. In some cases the original color name has shifted in meaning, as a historic name has been applied to a popular modern color. By convention, a contemporary mixture of pigments that replaces a historical pigment is indicated by calling the resulting color a hue, but manufacturers are not always careful in maintaining this distinction. The following examples illustrate the shifting nature of historic pigment names:

Titian used the historic pigment Vermilion to create the reds in the great fresco of Assunta, completed c. 1518.

- Indian Yellow was once produced by collecting the urine of cattle that had been fed only mango leaves. Dutch and Flemish painters of the 17th and 18th centuries favored it for its luminescent qualities, and often used it to represent sunlight. In the novel Girl with a *Pearl Earring*, Vermeer's patron remarks that Vermeer used "cow piss" to paint his wife. Since mango leaves are nutritionally inadequate for cattle, the practice of harvesting Indian Yellow was eventually declared to be inhumane. Modern hues of Indian Yellow are made from synthetic pigments.

- Ultramarine, originally the semi-precious stone lapis lazuli, has been replaced by an inexpensive modern synthetic pigment, French Ultramarine, manufactured from aluminium silicate with sulfur impurities. At the same time, Royal Blue, another name once given to tints produced from lapis lazuli, has evolved to signify a much lighter and brighter color, and is usually mixed from Phthalo Blue and titanium dioxide, or from inexpensive synthetic blue dyes. Since synthetic ultramarine is chemically identical with lapis lazuli, the "hue" designation is not used. French Blue, yet another historic name for ultramarine, was adopted by the textile and apparel industry as a color name in the 1990s. and was applied to a shade of blue that has nothing in common with the historic pigment ultramarine.

- Vermilion, a toxic mercury compound favored for its deep red-orange color by old master painters such as Titian, has been replaced in painters' palettes by various modern pigments, including cadmium reds. Although genuine Vermilion paint can still be purchased for fine arts and art conservation applications, few manufacturers make it, because of legal liability issues. Few artists buy it, because it has been superseded by modern pigments that are both less expensive and less toxic, as well as less reactive with other pigments. As a result, genuine Vermilion is almost unavailable. Modern vermilion colors are properly designated as Vermilion Hue to distinguish them from genuine Vermilion.

A pigment is a material that changes the color of reflected or transmitted light as the result of wavelength-selective absorption. This physical process differs from fluorescence, phosphorescence, and other forms of luminescence, in which a material emits light. Many materials selectively absorb certain wavelengths of light. Materials that humans have chosen and developed for use as pigments usually have special properties that make them ideal for coloring other materials. A pigment must have a high tinting strength relative to the materials it colors. It must be stable in solid form at ambient temperatures. For industrial applications, as well as in the arts, permanence and stability are desirable properties. Pigments that are not permanent are called fugitive. Fugitive pigments fade over time, or with exposure to light, while some eventually blacken. Pigments are used for coloring paint, ink, plastic, fabric, cosmetics, food and other materials. Most pigments used in manufacturing and the visual arts are dry colorants, usually ground into a fine powder. This powder is added to a binder (or vehicle), a relatively neutral or colorless material that suspends the pigment and gives the paint its adhesion.

- A distinction is usually made between a pigment, which is insoluble in its vehicle (resulting in a suspension), and a dye, which either is itself a liquid or is soluble in its vehicle (resulting in a solution). A colorant can act as either a pigment or a dye depending on the vehicle involved. In some cases, a pigment can be manufactured from a dye by precipitating a soluble dye with a metallic salt. The resulting pigment is called a lake pigment. The term biological pigment is used for all colored substances independent of their solubility.

- In 2006, around 7.4 million tons of inorganic, organic and special pigments were marketed worldwide. Asia has the highest rate on a quantity basis followed by Europe and North America. By 2020, revenues will have risen to approx US$ 34.2 billion. The global demand on pigments was roughly US$ 20.5 billion in 2009, around 1.5–2% up from the previous year. It is predicted to increase in a stable growth rate in the coming years. The worldwide sales are said to increase up to US$ 24.5 billion in 2015, and reach US$ 27.5 billion in 2018.

11.4   Manufacturing and industrial standards

Before the development of synthetic pigments, and the refinement of techniques for extracting mineral pigments, batches of color were often inconsistent. With the development of a modern color industry, manufacturers and professionals have cooperated to create international standards for identifying, producing, measuring, and testing colors.

First published in 1905, the Munsell color system became the foundation for a series of color models, providing objective methods for the measurement of color. The Munsell system describes a color in three dimensions, hue, value (lightness), and chroma (color purity), where chroma is the different from gray at a given hue and value.

By the middle years of the 20th century, standardized methods for pigment chemistry were available, part of an international movement to create such standards in industry. The International Organization for Standardization (ISO) develops technical standards for the manufacture of pigments and dyes. ISO standards define various industrial and chemical properties, and how to test for them. The principal ISO standards that relate to all pigments are as follows:

- ISO-787: General methods of test for pigments and extenders.
- ISO-8780: Methods of dispersion for assessment of dispersion characteristics.

Other ISO standards pertain to particular classes or categories of pigments, based on their chemical composition, such as ultramarine pigments, titanium dioxide, iron oxide pigments, and so forth.

Many manufacturers of paints, inks, textiles, plastics, and colors have voluntarily adopted the Colour Index International (CII) as a standard for identifying the pigments that they use in manufacturing particular colors. First published in 1925, and now published jointly on the web by the Society of Dyers and Colourists (United Kingdom) and the American Association of Textile Chemists and Colorists (USA), this index is recognized internationally as the authoritative reference on colorants. It encompasses more than 27,000 products under more than 13,000 generic color index names.

In the CII schema, each pigment has a generic index number that identifies it chemically, regardless of proprietary and historic names. For example, Phthalocyanine Blue BN has been known by a variety of generic and proprietary names since its discovery in the 1930s. In much of Europe, phthalocyanine blue is better known as Helio Blue, or by a proprietary name such as Winsor Blue. An American paint manufacturer, Grumbacher, registered an alternate spelling (Thalo Blue) as a trademark. Colour Index

International resolves all these conflicting historic, generic, and proprietary names so that manufacturers and consumers can identify the pigment (or dye) used in a particular color product. In the CII, all phthalocyanine blue pigments are designated by a generic color index number as either PB15 or PB16, short for pigment blue 15 and pigment blue 16; these two numbers reflect slight variations in molecular structure that produce a slightly more greenish or reddish blue.

Figure 11.1

A wide variety of wavelengths (colors) encounter a pigment. This pigment absorbs red and green light, but reflects blue, creating the color blue. Pigments appear the colors they are because they selectively reflect and absorb certain wavelengths of visible light. White light is a roughly equal mixture of the entire spectrum of visible light with a wavelength in a range from about 375 or 400 nanometers to about 760 or 780 nm. When this light encounters a pigment, parts of the spectrum are absorbed by the chemical bonds of conjugated systems and other components of the pigment. Some other wavelengths or parts of the spectrum are reflected or scattered. Most pigments are charge-transfer complexes, like transition metal compounds, with broad absorption bands that subtract most of the colors of the incident white light. The new reflected light spectrum creates the appearance of a color. Pigments, unlike fluorescent or phosphorescent substances, can only subtract wavelengths from the source light; never add new ones.

The appearance of pigments is intimately connected to the color of the source light. Sunlight has a high color temperature, and a fairly uniform spectrum, and is considered a standard for white light. Artificial light sources tend to have great peaks in some parts of their spectrum, and deep valleys in others. Viewed under these conditions, pigments will appear different colors.

Color spaces used to represent colors numerically must specify their light source. Lab color measurements, unless otherwise noted, assume that the measurement was taken under a D65 light source, or "Daylight 6500 K", which is roughly the color temperature of sunlight.

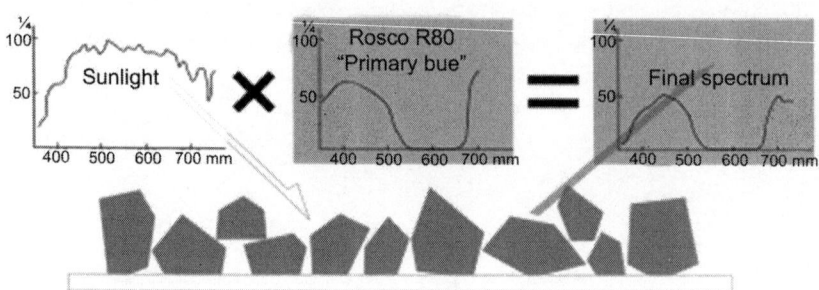

Figure 11.2

Sunlight encounters Rosco R80 "Primary Blue" pigment. The product of the source spectrum and the reflectance spectrum of the pigment results in the final spectrum, and the appearance of blue. Other properties of a color, such as its saturation or lightness, may be determined by the other substances that accompany pigments. Binders and fillers added to pure pigment chemicals also have their own reflection and absorption patterns, which can affect the final spectrum. Likewise, in pigment/binder mixtures, individual rays of light may not encounter pigment molecules, and may be reflected as is. These stray rays of source light contribute to a slightly less saturated color. Pure pigment allows very little white light to escape, producing a highly saturated color. A small quantity of pigment mixed with a lot of white binder, however, will appear desaturated and pale, due to the high quantity of escaping white light.

11.5   Differences between dyes and pigments

Dyes and pigments are substances that impart color to a material. The term colorant is often used for both dyes (also called dyestuffs) and pigments. The major difference between dyes and pigments is the solubility (the tendency to dissolve in a liquid). Dyes, also known as colorants in which the coloring matter is dissolved in liquid, are absorbed into the material to which they are applied. *Pigments*, on the other hand, consist of extremely fine particles of ground coloring matter suspended in liquid which forms a paint film that actually bonds to the surface it is applied to.

Pigments are *organic* or *inorganic*, *colored*, *white* or black materials that are practically insoluble in the medium in which they are dispersed. They are distinct particles, which gives the medium their color and opacity.

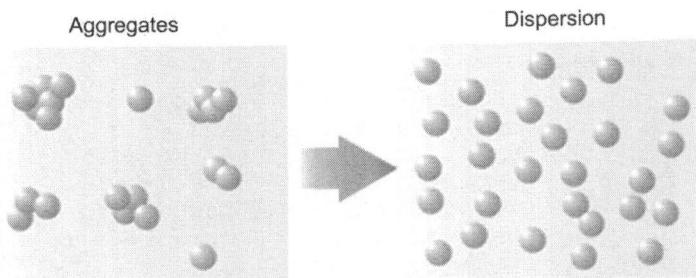

Figure 11.3 Pigments dispersion

The smallest units are called primary particles. The structure and shape of these particles depends on the crystallinity of the pigment. During the pigment production process, primary particles generally aggregate and generate agglomerates. During the dispersion of the pigment into the polymer, high shear is generally needed to break up these agglomerates (improved tinting strength). Pigments are thus required to resist dissolving in solvents that they may contact during application, otherwise problems such as "bleeding" and migration may occur. In addition, depending on the demands of the particular application, pigments are required to be resistant to light, weathering, heat and chemicals such as acids and alkalis.

The major differences between the dyes and pigments are highlighted below:

Points of difference	Dyes	Pigments
Solubility	They are soluble	Pigments are colourants that are insoluble in water and most of the solvents
Number	Available in Large number	Comparatively lesser in number
Product resistance	Lower as compared to pigments	Very high
Light fastness	Lower Dyes are very much vulnerable. Lights destroy colored objects by breaking open electronic bonding within the molecule	Traditionally pigments have been found to be more lightfast than dyes
Size	Dye molecules are comparatively smaller it's like comparing a football (pigment) to say a head of a pin (dye)	Pigment particles are about 1-2 microns in size. (1 micron =1/1000 meter). It means that the particles can be seen under a magnifying glass

Contd...

Contd...

Points of difference	Dyes	Pigments
Bonding	Taking the example of dyeing a wood surface, the dye and the substrate (wood) that is dyed are chemicals that have certain features called functional groups. At the level of molecules these groups serve as open pockets of electrostatic charges (+ or -). The functional group in dyes, serve as a method for attaching the dye to the wood	For example taking the example of a wood surface Pigment requires the help of a binder for gluing. As it is an inert substance which is merely suspended in a carrier/binder
Structure during the application process	During application process there is a temporary alteration in the structure of the dyes	During application, pigments have the capacity to retain particulate or crystalline structure
Imparting of colours	Dyes can only impart colour by selective absorption of the dyes	Pigments impart colours by either scattering of light or by selective absorption
Combustible properties	Taking the example of a Candle making process, if the candles are dyed it is easily combustible and can be applied throughout the candle	In the example of a candle making as pigments are colored particles, they tend to clog a wick when burned. This makes them undesirable for a candle if it is colored throughout and used for burning
Chemical Composition	Usually the dyes are organic (i.e. carbon-based) compounds	While pigments are normally inorganic compounds, often involving heavy toxic metals
Longevity factor	The dye based printing inks do not last as long as the pigment inks	In case of ink based printing prints made with pigments lasts longer
Printing on substrates	Compatible with almost all the substrates that needs to be dyed	Owing to the physical makeup of the pigment inks the range for suitable substrates are limited

Colorants are classified as either pigments or dyes. Pigments are practically insoluble in the medium in which they are incorporated. Dyes dissolve during application, losing their crystal or particulate structure in the process. The difference between pigments and dyes is therefore due to physical characteristics rather than chemical composition.

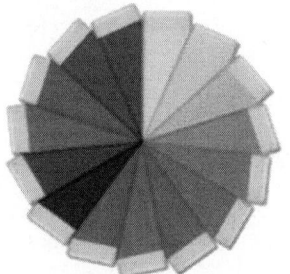

Figure 11.4

Dyes are used in the surface coating and printing industries, where a very high level of transparency is required for the production of colored metal foils and films, for flamboyant finishes in the paint industry and for wood stains, among numerous other applications. Thanks to advances in resin technology, dyes are now also being used in specific automotive finishes because of their color and transparency properties. Various solvents are suitable for these dyes. Dyes are also used for mass coloration of thermoplastics in a glass transition state at normal service temperatures. The dyes selected for the coloration of polymers are soluble within those substrates.

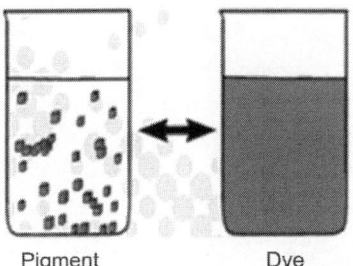

Pigment   Dye

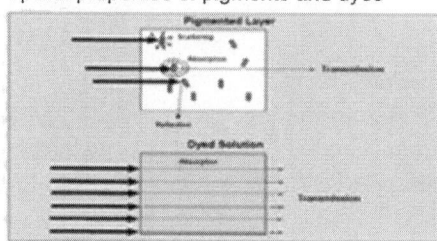

Optical properties of pigments and dyes

Figure 11.5

The dyes used fall into three main categories:
- Basic dyes
- Fat-soluble dyes
- Metal-complex dyes

11.6   Basic dyes

These cationic dyes are highly soluble in polar solvents such as alcohols, glycols and water. They are used by the printing ink industry with laking agents such as tannic acid to produce clean, bright shades. However, their poor light fastness limits their usefulness.

11.7   Fat-soluble dyes

These include nonionic, metal-free azo and anthraquinone dyes, which are highly soluble in less polar solvents, such as aromatic and aliphatic hydrocarbons. *Azo dyes* are mainly used for aliphatic/aromatic solvent-based wood stains and in the coloration of styrene polymers. *Anthraquinone dyes* are used much more widely in the coloration of plastics and fibers on account of their much wider range of resistance properties.

11.8   Metal-complex dyes

These are mainly anionic chromium and cobalt complexes of azo dyes. The cation is either a sodium ion or a substituted ammonium ion. Substituted soluble phthalocyanines also fall into this category. These dyes are normally soluble in alcohols, glycolethers, ketones and esters.

11.9   Dyes for wood coating

Designed for decorative and protective purposes, a wood stain is usually composed of a transparent or semi-transparent colorant solution. The solution penetrates the wood without hiding the grain, and indeed often enhancing it. The main requirements of wood colorants are transparency, high saturation/brilliance, good bleed and chemical resistance, good solvent compatibility and high light fastness. A number of factors influence staining results, including:

- *Substrate*: Wood type/color, cut of the wood, with/without pretreatment
- *Stain*: Concentration, application method (spraying, brushing, wiping …)
- *Clearcoat*: Over-lacquer resistance, color

Dyes can give brilliant shades and the excellent transparency that is a prime requirement in the wood coating industry, as it allows the wood grain to be seen.

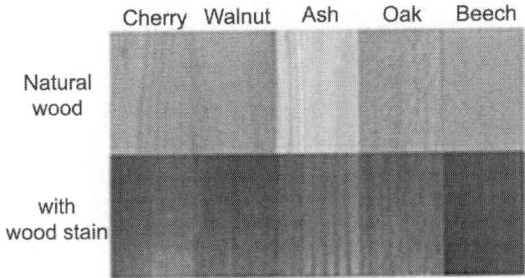

Cherry Walnut   Ash   Oak   Beech

Natural wood

with wood stain

2% ORA, SOL Red G wood stain without top
coat applied on different wood types

Figure 11.6

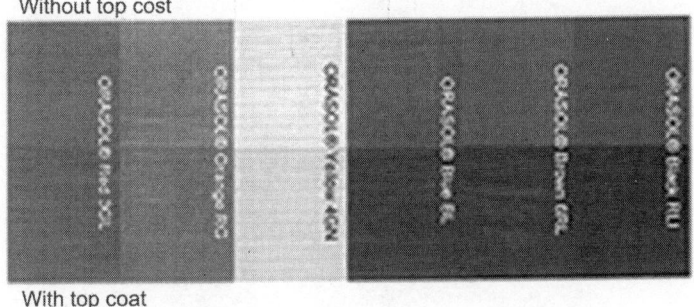

Without top cost

With top coat

Influence of top coat on dye appearance

Figure 11.7

Pigments are used in the coloration of paints, printing inks, ceramics and plastics. They can be used on a much wider variety of substances than dyes because they are not reliant on water solubility for their application. A pigment is a finely divided solid which is essentially insoluble in its application medium. In most cases the pigment is added to a liquid medium e.g. wet paint or a molten thermoplastic. The medium is then allowed to solidify by solvent evaporation or cooling and so the pigment molecules become fixed mechanically in the solid state.

The chromophores used in pigments are usually the same as those used in dyes but the pigments are large molecules and do not have solubilising groups. They contain groups that form intermolecular bonds that help to reduce solubilities. The larger the molecule, the more opaque the pigment.

Figure 11.8

In Figures 11.8 and 11.9, the red and yellow colorants are azo pigments (C.I. Pigment Red 57 and C.I. Pigment Yellow 13). The blue is the pigment copper phthalocyanine. [*Source*: By kind permission of BASF].

Figure 11.9

Organic pigments generally produce a higher intensity and brightness of colour than inorganic pigments such as chrome yellow (lead(II) chromate(VI)). Organic pigments exhibit a range of fastness properties that are dependent on the molecular structure and the nature of the intramolecular association in the solid state. An increase in the molecular size of a pigment generally decreases the solubility of the pigment. Also many pigments have the amide group (–NHCO–) incorporated which further decreases its solubility as the molecules are held together in large structures by hydrogen bonding (between the N–H group in one molecule and a C=O group in another).

Many organic pigments are based on azo chemistry and dominate the yellow, orange and red shade areas. An example of a simple monoazo pigment is C.I. Pigment Yellow 1:

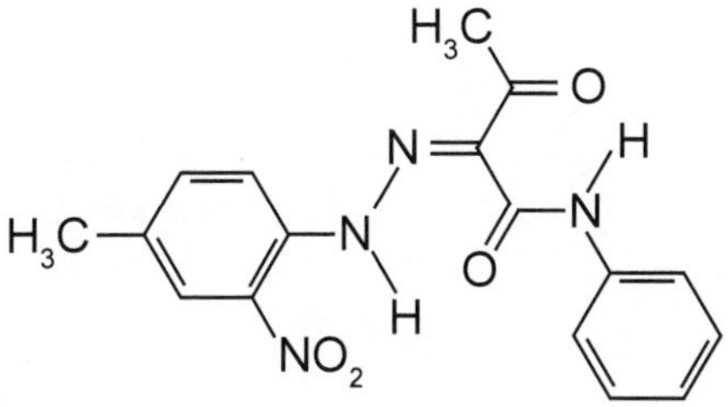

C.I. pigment yellow 1

Figure 11.20

This form is the ketohydrazone tautomer.

Copper phthalocyanines provide the majority of blue and green pigments. They are structurally complex but are relatively inexpensive to make. They provide excellent resistance to light, heat, acids and alkalis.

An example is C.I. Pigment Blue 15:

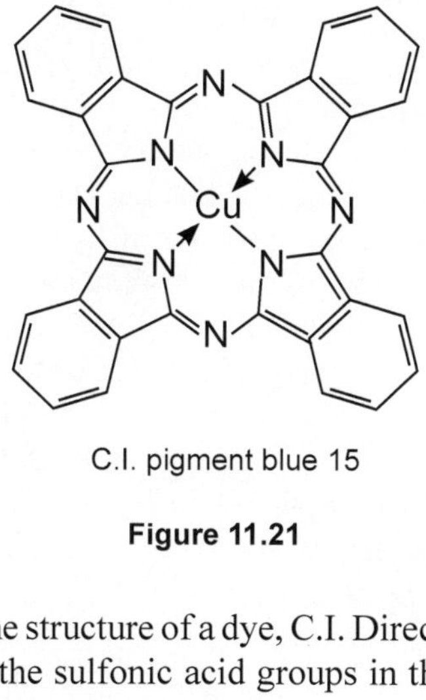

C.I. pigment blue 15

Figure 11.21

Earlier in the unit, the structure of a dye, C.I. Direct Blue 86, was displayed and it can be seen how the sulfonic acid groups in that structure transform a pigment into a dye.

Disperse dyes

12.1   Introduction

Before the First World War, almost all dyes were applied from solution in an aqueous dyebath to substrates such as cotton, wool, silk and other natural fibres. However, the introduction of a man-made fibre, cellulose acetate, with its inherent hydrophobic nature, created a situation where very few of the available dyes had affinity for the new fibre. Water-soluble anionic dyes had little substantivity for the fibre and the alkaline conditions required for the application of vat dyes brought about a loss in tensile strength and deterioration in fibre appearance due to the rapid hydrolysis of acetyl groups. The development of disperse dyes for dyeing secondary cellulose acetate fibres in the early 1920s was a major technological breakthrough, although their major use today is for the coloration of polyesters, the most important group of synthetic fibres. The first systematic study of dyes that was suitable for application to cellulose acetate by a direct dyeing process was carried out by Green. The presence of hydroxyl and amino groups, a low relative molecular mass and an almost neutral or basic character were found to be advantageous. As a result of these investigations, in 1922, Green and Saunders developed the Ionamine dyes (British Dystuffs Corporation) for application to acetate fibres. These water-soluble dyes were hydrolyzed in the aqueous dyebath to produce the sparingly soluble-free base in a very fine suspension that was then absorbed by the fibre. This discovery, that aqueous dispersions of almost water-insoluble dyes were highly suitable for the dyeing of secondary acetate, lead to the rapid development of other such dyes for dyeing cellulose acetate.

In 1923, aqueous dispersions of dyes were examined independently by the British Celanese Corporation and the British Dyestuffs Corporation and Ionamine dyes were superseded by ranges of disperse dyes, such as SRA (British Celanese Corporation) and Duranol (ICI), which were devoid of ionic solubilising groups. These sparingly water-soluble acetate dyes were applied to cellulose acetate in the form a fine aqueous dispersion. The advent of other man-made fibres, such as nylon in 1938 and acrylic in the early 1940s, both of which possess a significant hydrophobic nature, further increased the use of disperse dyes.

In 1922, Green and Saunders made one type of coloured azo compound, in which a solubilising group (for example, methyl sulphate, $-CH_2-SO_3H$) is attached to amino group. In dye bath, they are slowly hydrolyzed and produce azo compound and formaldehyde bi-sulphate. This free azo compound was capable of dyeing cellulose acetate fibres. This dye was named "ionamine". But this ion amine did not give satisfactory result in dyeing.

Later in 1924, Baddiley and Sheperdsen of the British Dyestuffs Corporation (Duranol Dyes) and by Ellis of the British Celanese Company (SRA Dyes) produced sulphoricinoleic acid (SRA) for dyeing acetate fibres. This SRA was used as dispersing agent. Later it was seen that SRA was capable of dyeing nylon, polyester, acrylic, etc. In 1953 this dye was named as "Disperse Dye".

It is known water-soluble or temporarily solubilized dyestuffs that are use for dyeing of natural fibres such as cotton, wool, silk, linen etc., which are essentially hydrophobic in nature and the dyeing is almost based on the aqueous system. Direct, acid, acid mordant, metal-complex, azoic, vat, solubilized vat and sulphur dyestuffs had been developed for colouring natural fibres. But when the hydrophobic fibres made their appearance, they presented a very serious problem, in that they could not be dyed with the then-available dyestuffs and the full exploitation of the textile potential of a new fibre will not be achieved unless it is dyeable. Commercial production of cellulose acetate fibres commenced in 1920, which has negligible substantivity for available dyes. The first disperse dyes developed in 1934 were comparatively still in their infancy and were used for the dyeing of secondary cellulose acetate. It was not until 1953 the current name was adopted. During the 1950s and 1960s the growth of polyester was tremendous and today 90% of disperse dyes are used in the coloration of polyester and its blends.

The work of Green and Saunders resulted in the development of Ionamines and the work by the British Dyestuff Corporation and British Celanese had given rise to the Duranol and SRA ranges, respectively. During the 1970s and 1980s disperse dyes were the fastest growing class of dye. There are literally hundreds of disperse dye manufacturers in the world and about 50 disperse dye ranges are available which together comprise more than 500 color entities. This is also evident from the number of disperse dyes being listed in the Color Index (C.I.). Third edition (1971) listed 554 disperse dyes, whereas the first revision (1975) and second revision (1982) listed 1106 and 1206 dyes, respectively.

Disperse dyes have experienced moderate patent activity over the last few years. Bright azo dyes containing heterocyclic systems have potential alternatives to fast anthraquinone dyes. Thiophene-based azo disperse dyes and ester-based disperse dyes are capable of clearing easily without the need for reduction clearing. The benzodifuranone derivatives (DyStar) with outstanding color fastness to thermomigration and wet treatments are developed. Alkali-stable disperse dyes (Dianix AD – DyStar) capable of being dyed in the pH range of 8.5 and 9.5 have been developed. Novel dyes continue to appear and the range of chromophore will increase within the next decade. Another area is the protection of disperse dyes from photodegradation in daylight.

The Society of Dyers and Colourists (UK) defines disperse dye as a substantially water-insoluble dye having substantivity for one or more hydrophobic fibres and usually applied from a fine aqueous dispersion. Disperse dyes are sold under various brand names, such as Foron (Clariant), Dianix (DyStar), Longsperse (Lonsen), Tulasteron (Atul), Terenix (Jaysynth), Terasil (Huntsman) and Coralene (Colourtex). Due to non-ionic nature, disperse dyes are volatile and the dye-vapour is strongly adsorbed by hydrophobic fibres. This is the basis for thermosol (heat-fixation) dyeing and heat transfer printing processes. The first transfer paper was produced on commercial scale in 1968 and in 1975 transfer prints became highly fashionable. Disperse dyes are mainly used for acetate and polyester fibres. Polyester fibres are dyed at high temperature. The dyed and printed polyester materials are often subjected to heat treatment. Hence dyes of greater molecular size having good sublimation and light fastness and giving good colour yield on polyester are desirable.

On nylon, disperse dyes can cover barre effect (structural differences) due to variation in end amino groups. However, due to limited fastness properties, dyeing is restricted to light and medium shades on nylon hosiery and carpets. On acrylic fibres, light fastness of disperse dyes is good, but limited colour build-up restricts the dyeing to light shades only.

12.2   Characteristics of disperse dyes

1. Disperse dyes are low molecular weight substances, mostly derived from azo, anthraquinone and diphenylamine.
2. Disperse dyes are crystalline materials of high melting point (>150°C). They are milled with dispersing agents to produce stable dispersions in dyebath and 0.5 to 2 micron in particle size. As they are applied in the form of very fine aqueous dispersions, both particle size and dispersion stability are extremely important. Ideally, disperse dyes should disperse extremely rapidly when added to water and give a stable dispersion of very fine and uniform particle size.

3. These dyes are non-ionic in nature and have relatively low solubility in water.

4. They produce a stable dispersion in the dyebath.

5. These dyes usually have NO_2, CN, OH, halogen and amino groups.

6. Due to low molecular weight of disperse dyes, they facilitate their entry and diffusion into the crystalline polyester fibre. The higher the molecular weight of dyes, the slower is the diffusion in the fibre.

7. Disperse dyes are marketed in both powder and liquid forms. The powder form contains considerable amount of dispersing agent, e.g. a naphthalene sulphonic acid-formaldehyde condensate and possibly a wetting agent. The liquid dyes are concentrated and free-flowing aqueous dispersions. The advantages claimed for the liquid brand are freed from dusting, easy preparation of dyebath and padding liquor and suitability for automatic metering systems called autodispensing. They contain lower amounts of dispersing agents and give higher colour-yield and brighter dyeing in pad-bake process.

8. The colour yield, brightness and shade of disperse dyes remain unaffected in the presence of hard water of 50 °C English hardness. A number of disperse dyes are susceptible to copper or iron and form a blue-violet complex with them. This may cause change of shade or staining. So in such cases it is advisable to add sequestering agents (1 gpl).

9. Disperse dyes show thermal migration, e.g. migration of dye from core to surface while drying of dyed goods at 170 °C or above. Migration is more in case of disperse dyes of high molecular weight having high sublimation fastness.

Such type of complain comes from Bhilwara and Mumbai market where the dyed p/v fabric is processed at higher temperature. To overcome such problems, the fibre-dyed fabric must be scoured properly before finishing (to remove any spinning or knitting oils) and finishing agents may also cause thermal migration of disperse dyes. In the finishing process non-ionic emulsifiers and finishing agents should be avoided. Instead use anionic and cationic agents which are less harmful.

12.3   Classification of disperse dyes

The important developments with regard to individual dye structures are conveniently classified by reference to the chromophoric groups like:

(a) Monoazo (50%)

(b) Anthraquinone (25%)

(c) Disazo (10%)

The rest comprise of styrl (3%), methine (3%), aroylenebenzimidazole (3%), quinonaphthalone (3%), nitro (1%), aminonaphthlimide (1%) and naphtha-quinoneimine (1%).

The monazo dyes had relatively poor affinity. But the disazo dyes have good light fastness, and sometimes polar groups are introduced to increase the sublimation fastness. They are mainly yellows and oranges. Azo-based dyes are dischargeable dyes so commonly used in printing.

Some commercial Amino azobenzene dyes are:

1. C.I. Disperse Orange 25

2. C.I. Disperse Orange 30

3. C.I. Disperse Red 90

4. C.I. Disperse Blue 183

5. C.I. Disperse Blue 79

Some typical anthraquinone disperse dyes are bright red to blue dye and their stability towards reduction and hydrolysis are good. The main disadvantages of these dyes are that they are somewhat weaker than many azo dyes and several intermediate stages are required in their production increasing the cost.

Improvement in light fastness of about 7 can be obtained by incorporating electron-attracting groups. Sublimation fastness can be improved by incorporating polar groups or by increasing molecular size. Anthraquinone-base dyes are non-dischargeable in nature so used for plain dyeing purpose and over printing purpose.

The dyes are as follows:

1. C.I. Disperse Violet 27

2. C.I. Disperse Blue 73

3. C.I. Disperse Blue 60

4. C.I. Disperse Red 15

5. C.I. Disperse Blue 3

BASF characterized their Palanil disperse dyes by diffusion number, which is based on the degree of penetration of the dye into polyester film under defined dyeing conditions. The number is important for assessing some important properties of disperse dyes, notably its combinability and suitability for a particular application. The larger the molecular size of the dye, slower will be the rate of diffusion and high energy will be required for its application. The sublimation fastness for such dye will also be higher. Therefore, disperse dyes may be classified into three groups:

1. Low energy disperse dyes
2. Medium energy disperse dyes
3. High energy disperse dyes

The first type of disperse dye forms at lower temperature with addition of trichlorobenzene base carrier. These dyes have low sublimation fastness properties but due to their low molecular weight these dyes are level dyeing dyes. High energy disperse dyes are dyed either in HTHP or Thermosol method. These dyes possess high molecular weight (>500), so it is called high sublimation fastness dyes though these class of disperse dyes possess poor leveling properties due to their low migration in dyeing.

Medium energy-type dyes stand between these two groups. For example, Clariant classify their disperse dyes in same three categories with following headings depending on the leveling character, molecular size and sublimation fastness of the dyes like:

1. *Foron E brand disperse dyes* – Low energy disperse dyes, excellent leveling properties, low sublimation fast dyes, suitable for piece dyeing of polyester/blends.

2. *Foron SE brand disperse dyes* – Medium energy disperse dyes, medium sublimation fast dyes, good leveling dyes, suitable for general applications on polyester/ blends.

3. *Foron S brand disperse dyes* – High energy disperse dyes, high sublimation fast dyes, suitable for print/pad/ high temperature or high pressure steam fixation. These are also recommended for thermosol dyeing of polyester/blends.

The committee on disperse dyes of the Society of Dyers and Colourists (U.K.) formulated tests for classifying disperse dyes on cellulose acetate on an A-E scale. Class-A dyes have best migration properties, widest range of application temperature, best colour build-up and fastest in dyeing rate. Class-E is the worst in the above properties and the dyeing rate is slowest. On the basis of exhaust dyeing of polyester at high temperature (130 °C) which covered critical dyeing temperature, migration, build-up and diffusion rate is expressed on A-D or A-E scale, where A represents the most satisfactory in relation to the above properties and D or E represents least satisfactory for the respective scale. Lastly disperse dyes may also be rated according to their penetrating or diffusion power. Latest the disperse dyes are classified as per special application like.

1. *Rapid dyes* – Known as RD dyes. Here the temperature gradient can go up to 3–4°C/min. without affecting the quality of dyeing.

2. *High light fastness dyes* – Used for automotive fabrics.

3. *High wash fastness dyes* – Required for sportswear and swimwear application.

4. *Alkali stable dyes* – Suitable for one bath coloration of polyester / cellulosic blends with reactive dyes. Also used for oligomer-free dyeing of polyester fibre.

5. *Fluorescent dyes* – Suitable for coloring high visibility clothing.

With the rapid development of polyester microfibres, there is an urgent need to develop suitable disperse dyes. Although many works have been done in this field, most of the dyes are selected from the conventional disperse dyes, such as C.I. Disperse Blue 257, C.I. Disperse Red 135. A series of disperse dyes bearing ether groups have been synthesized and they have been found to have good fastness on polyester microfibres.

Presently, dyeing properties are studied of capsules containing disperse dyes into polyester fabrics. It is known as microcapsule dyes (MC-dye). It was prepared by encapsulating both disperse dyes and ferromagnetic substance with acrylate resin. MC-dye deposited polyester fabric was dyed by heat treatment ranging 170–210°C. Longsperse disperse dyes are manufactured by Longsheng, China, which is the biggest disperse dyestuff manufacturer in the world. The term "disperse dye" have been applied to the organic colouring substances which are free from ionizing groups, are of low water solubility and are suitable for dyeing hydrophobic fibres. Disperse dyes have substantivity for one or more hydrophobic fibres e.g. cellulose acetate, nylon, polyester, acrylic and other synthetic fibres.

The negative charge on the surface of hydrophobic fibres like polyester cannot be reduced by any means, so non-ionic dyes like disperse dyes are used which are not influenced by that surface charge.

12.4   Properties of disperse dyes

• Disperse dyes are nonionic dyes. So they are free from ionizing group.

• They are readymade dyes and are insoluble in water or have very low water solubility.

• They are organic colouring substances which are suitable for dyeing hydrophobic fibres.

- Disperse dyes are used for dyeing manmade cellulose ester and synthetic fibres especially acetate and polyester fibres and sometimes nylon and acrylic fibres.
- Carrier or dispersing agents are required for dyeing with disperse dyes.
- Disperse dyes have fair to good light fastness with rating about 4–5.
- They melt at above 150 °C.
- They are crystalline in nature which are ground with dispersing agents to produce particles of 0.5–4 millimicron in size and which produce a stable dispersion in the dye bath.

12.5   Classification of disperse dyes

According to chemical structure:
1. Nitro dyes (Nitrodiphenyl derivatives)
2. Amino ketone dyes
3. Anthraquinonoid dyes
4. Mono azo dyes
5. Di-azo dyes

These dyes are essentially hydrophobic and are almost insoluble in water. However, they have affinity for hydrophobic fibres, for example polyesters, and are applied as very fine dispersions in water. Most disperse dyes are azo compounds and can give colours across the spectrum. Some are anthraquinone-based dyes for reds, violets, blues and greens. Polyester fibres can be dyed at 400 K under pressure, allowing the use of larger molecular size dye structures which achieve better fastness, for example:

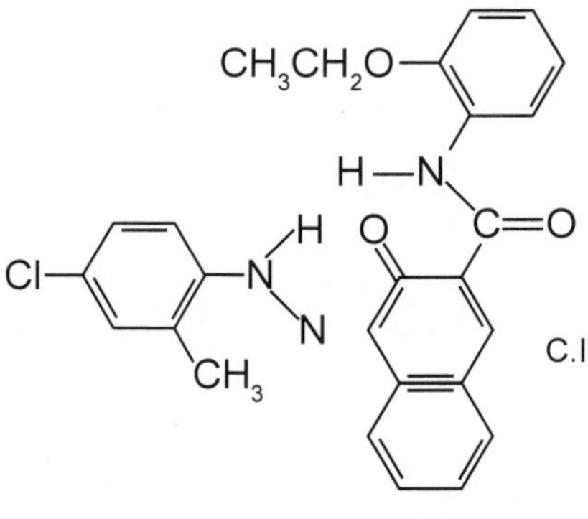

C.I. disperse red 220

Figure 12.1

The structure shown is the ketohydrazone tautomer.

12.6   Manufacturing of disperse dyes

Disperse dyes are essentially low molecular weight derivatives of azo, anthroquinone and other compounds. They are essentially non-ionic in nature and contain aromatic or aliphatic amino, mono- and di-substituted amino and hydroxyl groups in their molecular structure.

The violet, blue and green anthraquinone dyes used on polyester are commonly synthesized from 1,4-diamino anthraquinone, employing two different alkyl or aryl groups. The yellow to red shades are commonly obtained by means of azo dyes.

12.6.1   Dispersion

Dispersion of disperse dyes can be achieved as follows:
  (a) By using powerful dispersing agent or
  (b) By attaching alkanol, carboxyl amide and other groups to the dye molecule.

During the manufacture, commercial disperse dyes are milled with a dispersing agent. The sodium salt of a cresolnaphthalene sulphonic acid formaldehyde condensate, thus enabling the particle side to be reduced to 2–4 millimicron, which is maximum size for satisfactory dispersion.

12.6.2   Levelling agents

Selected non-ionic surface-active agents can also be used to increase the solubility of disperse dyes in water and increased rate of migration, leveling and fibre penetration.

12.6.3   Disperse Yellow 13 (Disperse Yellow 8 g)

In an autoclave charge 20 kg 3-Bromobenzanthrone, 40 litre methanolic NaoH solution 10% (W/V), and 380 litre methanol. Heat to 130–135°C. Maintain at 130–135°C for 25 hours, under stirring. Cool to 25–30°C. Charge is diluted with 500 litres of water. The pH is alkaline. Stir for 1 hour filter, wash with water, till free from alkali. Suck the product and dry at 80°C.

12.6.4   Acid pasting

In an acid pasting vessel charge 170 kg of 90% sulphuric acid. Cool to 8–10°C. Charge 17 kg of 3-methoxy benzathrone slowly in 1 hour at 8–10°C. Stir for 2 hours at 8–10°C to dissolve 3-methoxy benzathrone. Drown the charge slowly in 500 litres water at 10°C. Temperature after drowning should be 18–20°C. Drowning should be carried out slowly. Filter, wash with water till acid free, suck.

12.6.5   Dispersion

Add anionic emulsifier and small quantity of soda ash to keep the product just alkaline. Dry at 50–60 °C.

12.7   Application methods of disperse dyes

1. *Method N*: Normal dyeing method. Dyeing temperature is 80–100°C.
2. *Normal NC method*: Method of dyeing at normal temperature with carriers. Dyeing temperature 80–100 °C.
3. *Method HT*: High temperature dyeing method. Dyeing temperature 105–140 °C.
4. *Method T*: Thermasol dyeing method. Dyeing temperature 180–220 °C, continuous method of dyeing.
5. *Pad roll method*: Semi-continuous dyeing method.
6. *Pad steam method*: Continuous dyeing method.

12.8   Dyeing mechanism of disperse dye

The dyeing of hydrophobic fibres like polyester fibres with disperse dyes may be considered as a process of dye transfer from liquid solvent (water) to a solid organic solvent (fibre). Disperse dyes are added to water with a surface-active agent to form an aqueous dispersion. The insolubility of disperse dyes enables them to leave the dye liquor as they are more substantive to the organic fibre than to the inorganic dye liquor. The application of heat to the dye liquor increases the energy of dye molecules and accelerates the dyeing of textile fibres.

Heating of dye liquor swells the fibre to some extent and assists the dye to penetrate the fibre polymer system. Thus the dye molecule takes its place in the amorphous regions of the fibre. Once taking place within the fibre polymer system, the dye molecules are held by hydrogen bonds and Vander Waals' force.

The dyeing is considered to take place in the following simultaneous steps:

Diffusion of dye in solid phase into water by breaking up into individual molecules. This diffusion depends on dispersibility and solubility of dyestuff and is aided by the presence of dispersing agents and increasing temperature.

Adsorption of the dissolved dye from the solution onto the fibre surface. This dyestuff adsorption by fibre surface is influenced by the solubility of the dye in the dye bath and that in the fibre.

Diffusion of the adsorbed dye from the fibre surface into the interior of the fibre substance towards the centre. In normal condition, the adsorption rate is always higher than the diffusion rate. And this is the governing step of dyeing.

When equilibrium dyeing is reached, the following equilibria are also established:

1. Dye dispersed in the bath
2. Dye dissolved in the bath
3. Dye dissolved in the bath
4. Dye adsorbed on the fibre
5. Dye adsorbed on the fibre
6. Dye diffused in the fibre

12.9   Effect of various conditions on disperse dyeing

Effect of temperature – In case of dyeing with disperse dye, temperature plays an important role. For the swelling of fibre, temperature above 100°C is required if high temperature dyeing method is applied. Again in case of carrier dyeing method, this swelling occurs at 85–90°C. If it is kept for more time, then dye sublimation and loss of fabric strength may occur.

Effect of pH – For disperse dyeing, the dye bath should be acidic and pH should be in between 4.5 and 5.5. For maintaining this pH, generally acetic acid is used. At this pH, dye exhaustion is satisfactory. During colour development, correct pH should be maintained otherwise fastness will be inferior and colour will be unstable.

12.10   Carrier dyeing method

Procedure

- At first, a paste of dye and dispersing agent is prepared and then water is added to it.

- Dye bath is kept at 60 °C temperature and all the chemicals along with the material are added to it. Then the bath is kept for 15 min without raising the temperature.
- pH of bath is controlled by acetic acid at 4–5.5.
- Now temperature of dye bath is raised to 90 °C and at that temperature the bath is kept for 60 min.
- Then temperature is lowered to 60 °C and resist and reduction cleaning is done if required. Reduction cleaning is done only to improve the wash fastness.
- Material is again rinsed well after reduction cleaning and then dried.

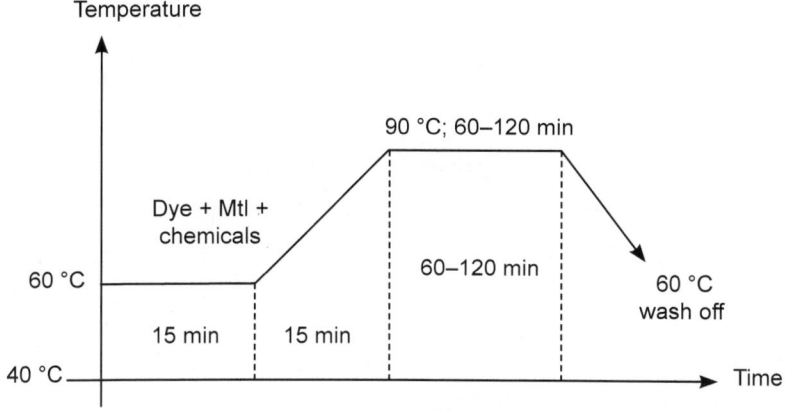

Figure 12.2 Dyeing Curve

12.11   High temperature dyeing method

Procedure

- At first a paste of dye and dispersing agent is prepared and water is added to it.
- PH is controlled by adding acetic acid.
- This condition is kept for 15 minutes at temperature 60 °C.
- Then the dye bath temperature is raised to 130 °C and this temperature is maintained for 1 hour. Within this time, dye is diffused in dye bath, adsorbed by the fibre and thus required shade is obtained.
- The dye bath is cooled as early as possible after dyeing at 60 °C.
- The fabric is hot rinsed and reduction cleaning is done if required.
- Then the fabric is finally rinsed and dried.

Temperature

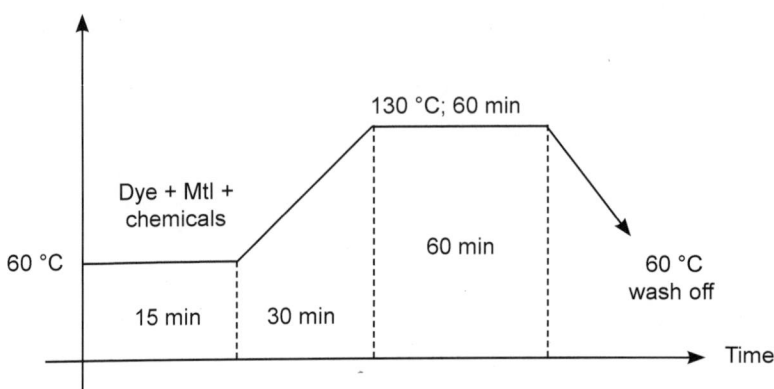

Figure 12.3 Dyeing Curve

12.12   Dyeing of polyester fabric in thermasol dyeing method

Thermasol dyeing method is continuous methods of dyeing with disperse dye. Here dyeing is performed at high temperature like 180–220 °C in a close vessel. Here time of dyeing should be maintained very carefully to get required shade and to retain required fabric strength.

Sequence: Pading — Drying — Thermofixing — After-treatment

Procedure:

1. At first the fabric is padded with dye solution using above recipe in a three bowl padding mangle.
2. Then the fabric is dried at 100 °C temperature in dryer. For dyeing, infra red drying method is an ideal method by which water is evaporated from fabric in vapor form. This eliminates the migration of dye particles.
3. Then the fabric is passed through thermasol unit where thermo fixing is done at about 205 °C temp for 60–90 seconds depending on type of fibre, dye and depth of shade. In thermasol process about 75–90% dye is fixed on fabric.
4. After thermo fixing, the unfixed dyes are washed off along with thickener and other chemicals by warm water.

Then soap wash or reduction cleaning is done if required. And finally the fabric is washed.

12.13   New developments in disperse dyes

New generation of specially designed disperse dyes to meet the highest automotive light fastness requirements for multiple cycles and 488 kJ. These dyes show excellent sublimation fastness and "on-tone" fading. The field of application are automotive textiles. It involves characterizing the chemistry associated with the light-induced (photo) degradation of azo, anthraquinone, and nitrodiphenylamine disperse dyes in an ester environment. There are built-in photostabilizer groups for enhancing disperse dye lightfastness (resistance to light-induced fading). In this regard, dyed polyester films were exposed to sunlight from two regions in the United States and to the artificial light of an Atlas Weatherometer. The results of the different exposures indicate that artificial light exposures were far more damaging to the polymer host than natural light. It was also apparent that a significant level of dye fading could be attributed to substrate degradation.

The use of polyester fabrics in automotive upholstery has created a need for disperse dyes of exceptionally high sublimation and light fastness to withstand extended exposure to light and heat. Thus, in an attempt to find new disperse dyes of high fastness to sublimation and light, the synthesis and evaluation of their fastness properties on polyester fabrics and nylon 66 fabrics was done. The synthesis of the dyes was carried out by diazotisation of the 2-amino-5-mercapto-1,3,4-thiadiazole in an acid medium and coupling with 3-chloroaniline to afford an intermediate. This intermediate was also subjected to diazotisation and coupling reactions with various couplers to give disazo disperse dyes. All the dyes exhibited outstanding washing, sublimation, perspiration and good to very good light fastness ratings on polyester and nylon 66 fabrics.

12.14   Stripping of disperse dyes

Stripping of disperse dyed material can be accomplished by treating goods:

(a) In a blank bath containing non-ionic leveling agent at 130 °C for 45 min to 60 min.

(b) For chemical destruction of the dye use 1–2 gpl of Caustic Soda and a reducing agent.

(c) Mild acidic treatment with zinc sulfoxalate formaldehyde or sodium chlorite can reduce the shade to almost off-white background.

Partial stripping of the colour of PET materials dyed with disperse dyes is usually possible by treatment with a solution of dyeing carrier or retarding agent at high temperature under pressure. Oxidative and reductive stripping is also possible but is likely to involve some undesirable effects upon the fabric handle or appearance. Prolonged treatment of polyester materials with alkaline solutions causes surface hydrolysis of ester groups and loss of weight. Once the surface has been degraded, it is difficult to obtain the originally anticipated appearance.

Reactive dyes

13.1 Introduction

It took 100 years of dye chemistry to produce reactive dyes following Perkin's discovery of the first synthetic dye, Mauvein or Mauve in 1856. This obvious and uncomplicated principle went unnoticed for so long. The reason for this neglect is that most dye chemists believed cellulose to be relatively inert. More attention for developing reactive dyes was for wool fibre, which is more chemically active, though the need and the potentiality of the change of technology were not as great as in the case of cellulosic fibres. In the 1920s Ciba had recognized that monochlorotriazine dyes had good wet fastness on wool. IG Farben marketed similar dyes in 1930s. In 1949 Hoechst patented the first precursors of vinyl sulphone dyes and two of these were introduced as Remalan dyes for wool in 1952. Though reactive dyes were originally developed for wool, the commercial exploitation of reactive dyes for wool started after the introduction of Lanasol (Ciba) dyes.

The most important event in the history of reactive dyes was introduction of reactive dyes for cellulose by ICI in 1954. Rattee and Stephan of ICI claimed that when water-soluble dye containing a dichlorotriazine (DCT) group is applied to cellulose from a neutral bath and then increasing the pH value, a covalent bond is formed between a triazine carbon atom and an oxygen atom of a cellulose hydroxyl group. The introduction of reactive dyes (Procion Yellow R, Brilliant Red 2B and Blue 3G) by ICI in 1956 is an important landmark in the history of synthetic dyes. Subsequently, the range of reactive dyes steadily increased as other countries began manufacturing them. Immediately after introduction of Procion dyes, all major dyestuff manufacturers began investigation for newer reactive dyes.

Ciba manufactured monochlorotriazine dyes (MCT) or aminochlorotriazine, which can be applied on cellulose at higher temperature and pH. Ciba named the dyes as Cibacron dyes, while the same dyes were introduced by ICI in the name of Procion H (H for hot), both in 1957. Later Ciba introduced monofluorotriazine (MFT) dyes in the name of Cibacron F. Hoechst introduced Remalan dyes based on vinylsulphone (VS) for wool in 1949, claimed to be suitable for cellulose too. However, they introduced Remazol reactive dyes exclusive for cellulose containing sulphuric acid ester of B-hydroxyethylsulphone in 1958. In subsequent years, various dyestuff manufacturers introduced further reactive groups.

Though reactive dyes were originally developed for wool, the commercial exploitation of reactive dyes for wool started after introduction of Lanasol (Ciba) dyes based on alphabromoacrylamide (–NHCO–C (Br) = CH$_2$) by Ciba. These dyes, as well as difluorochloropyrimidine dyes, e.g. Drimalan (Clariant) and Verofix (Bayer) dyes, have high fixation ratio.

Reactive dyes are the youngest and most important dye-class for cellulosic materials. The reactive dyes offer a wide range of dyes with varying shades, fastness, cost with high brilliancy, easy applicability and reproducibility. However good pretreatment of the material is a prerequisite. The colour yield and brilliancy of shades are enhanced significantly by mercerization. Unlike vat dyes, most of the reactive dyes are not fast to chlorine bleaching. Reactive dyes are water-soluble anionic dyes. Reactive dyes are purposefully designed to have low substantivity in order to facilitate removal of hydrolysed dye by simple soaping and rinsing.

13.2   Components of reactive dyes

The general formula of reactive dyes containing reactive system can be represented as:

S–F–T–X

Where,

S = Solubilising groups (such as SO$_3$Na or COONa or combination of both) which impart solubility.

F = Chromophoric group usually an azo, metal-complex azo, anthraquinone, triphenyl dioxazines and formazan molecules or phthalocynanine residue. The chromophore is responsible for the colour, affinity and diffusion of dye.

T = Bridging group which attach the reactive system X to the chromogen F, this group is usually –NH, –O–, –NHCO–, –OCH$_3$–, –SO$_3$–, etc.

X = Reactive system or group, which reacts chemically with the functional group of the fibre with the formation of covalent bond between the dye and the fibre.

Many fibre reactive dye systems have been developed over the last 50 years, particularly for cellulosic fibres. Development of new reactive systems is to obtain decreased substantivity, increased resistance to hydrolysis and improved fixation. Substantivity can be decreased by adding solubilising groups and disturbing the co-planarity of the reactive dye molecule by adding substituting group ortho to the point of attachment of the planar reactive or N-substitution of nitrogen atom.

Conventional reactive dyes for cellulosic fibres suffer an obvious drawback in that only 70% of the dye is fixed onto the fibre; the remainder of the dye undergoes hydrolysis in the dye bath. There has been a continuous search for dyes that will exhibit high fixation efficiency. There are two possible methods to accomplish the goal:

1. Use a reactive group that is not easily hydrolyzed.

2. Incorporate two or more reactive groups in a dye molecule to increase the chance of fixation.

13.3   Types of reactive dyes

Mainly reactive colours are of five types. Reactive colours form a chemical linkage with the cellulosic fibres in the presence of alkalis and therefore are characterized by their excellent fastness. Reactive colours are easy to use and applicable extensively in various fields such as exhaust dyeing, continuous dyeing and printing, etc.

1. *Bi-functioal dyes (also known commonly as ME dyes)* – These are reactive dyes containing monochlorotriazinyl and vinyl sulphone group as a reactive radical and useful for dyeing of cellulosic fibres. These colours are suitable for exhaust dyeing (60°C), continuous dyeing and printing.

2. *Vinyl sulphone dyes – These are* reactive dyes containing vinyl sulphone groups as a reactive radical. These are suitable for exhaust dyeing (60°C), continuous dyeing and printing.

3. *H-E dyes* – These are reactive dyes containing monochlorotriazinyl group as a reactive radical and high fixation on dyeing fabric blends on polyester/cotton. These colours are suitable for exhaust dyeing (80°C) of medium/heavy colours as high colour yield.

4. *H-dyes (Hot brand)* – These are reactive dyes containing monochlorotriazinyl group as a reactive radical. These dyes are useful for printing of cellulosic fibres and viscose rayon. They are applied to steaming at 100–105°C.

5. *M-dyes (Cold brand)* – These are reactive dyes containing dichloro triazinyl group as reactive group. These colours are suitable for exhaust dyeing and also can be applied to cellulosic fibre at room temperature 35°C.

First came the cyanuric chloride reactive dyes; then the vinyl sulphone and the tetrachloro pyrimidine reactive dyes. The cyanuric chloride and the tetrachloro pyrimidine dye cotton mainly whereas the vinyl sulphone dyes acetates, polyamides and wool. The Remazol dyes (vinyl sulphone) form an ether bondage with the fibre whereas the Cibacron and Procion form an ester linkage. It is found that about 50% of the reactive dyes have a phthalocyanine groups. Now light fastness is an important requirement of export-oriented textile substrate. They belong to fluorine chemistry and made by Ciba known as Cibacron F and FN dyes. It is not possible to manufacture in India.

13.4   Dyeing methods

Dyeing of reactive dyes, as in the case of many other classes of dyes, can be carried out by batch dyeing process and continuous dyeing process.

1. *Dyeing with M-brand dyes* – These are cyanuric chloride-based reactive dyes, highly reactive in nature and generally employed under exhaust method at 30–40°C and pad-batch method. These dyes possess brilliancy in shades with high colour fastness to washing and light. These fibre-reactive dyes are easy to dyeing with high consistency of shades.

Following are the pad-batch method:

(a) Pad-batch method (soda ash with M brand dyes or soda ash + bicarbonate with M brand dyes etc.)

(b) Pad-dry-thermofix method (bicarbonate + M dyes)

(c) Pad-dry-wash-method (bicarbonate + M dyes)

(d) Pad-steam method (bicarbonate + M dyes)

2. *Dyeing with H-brand dyes* – These H-brand reactive dyes contain brilliant shades of high fastness on cellulosic fabric. These dyes are less reactive in nature so require more severe conditions of fixation.

These dyes are suitable for :

(a) Exhaust method

(b) Pad-batch method

(c) Semi-continuous methods

(d) Continuous methods

3. *Dyeing with ME reactive dyes* – These reactive dyes are most suitable at low temperature at 60–65°C for high exhaust purpose and pad-batch method. These are structurally different dyes from conventional dyes containing higher grade wet fastness, i.e. perspiration fastness, light fastness, washing fastness, dry and wet crocking fastness, chlorine and peroxide washing fastness, etc. High degree of exhaustion and fixation rate is another merit point to

accelerate excellent leveling properties. They have better alkali stability and reproductability behaviour.

4. *Dyeing with HE reactive dyes* – These dyes are identified for its extraordinary wet colour fastness along with cost saving compared to conventional reactive dyes. These dyes are suitable for cellulose-blended fabric too. It facilitates quick rinsing and soaping to save time factor. They yield batch-to-batch consistency. Dyeing can be controlled with salt and temperature before alkali. These dyes are less sensitive to Glauber salt, time and temperature besides their level dyeing results. These are specially made for exhaust dyeing style at 80°C to 90°C temperature.

5. *Dyeing with VS reactive dyes* – These dyes chemically react with the hydroxyl group of cellulose and form firm covalent linkage in presence of alkali. These dyes are employed by exhaust and pad batch process. Such dyes possess poor affinity for cellulosic fibre in absence of salt and alkali. For this reason these dyes are most suitable for pad-batch method due to their high solubility even in presence of alkali. Similarly, substantivity can be increased by addition of Glauber salt or common salt and alkali for making the dyestuff well suitable for all type of conventional dyeing machines.

A dye, which is capable of reacting chemically with a substrate to form a covalent dye substrate linkage, is known as reactive dye. Here the dye contains a reactive group and this reactive group makes covalent bond with the fibre polymer and act as an integral part of fibre. This covalent bond is formed between the dye molecules and the terminal –OH (hydroxyl) group of cellulosic fibres on between the dye molecules and the terminal –NH2 (amino) group of polyamide or wool fibres.

Reactive dyes first appeared commercially in 1956, after their invention in1954 by Rattee & Stepheness at the Imperial Chemical Industry (ICI), Dyestuffs Division site in Bleckley, Manchester, UK. Reactive dyes form a covalent bond between the dye and fiber. It contain a reactive group (often trichlorotriazine), either a haloheterocycle or an activated double bond, that, when applied to a fibre in an alkaline dye bath, forms a chemical bond with an hydroxyl group on the cellulosic fibre. In a reactive dye a chromophore contains a substituent that is activated and allowed to directly react to the surface of the substrate.

Reactive dyeing is now the most important method for the coloration of cellulosic fibres. Reactive dyes can also be applied on wool and nylon; in the latter case they are applied under weakly acidic conditions. Reactive dyes have a low utilization degree compared to other types of dyestuff, since the functional group also bonds to water, creating hydrolysis.

13.5   Reactive dyes categorized by functional group

Functions	Fixation	Temperature	Included in brands
Monochlorotriazine	Haloheterocycle	80°	Basilen E & PCibacron EProcion H,HE
Monofluorochlorotriazine	Haloheterocycle	40°	Cibacron F & C
Dichlorotriazine	Haloheterocycle	30°	Basilen MProcion MX
Difluorochloropyrimidine	Haloheterocycle	40°	Levafix EADrimarene K & R
Dichloroquinoxaline	Haloheterocycle	40°	Levafix E
Trichloropyrimidine	Haloheterocycle	80–98°	Drimarene X & ZCibacron T
Vinyl sulfone	Activated double bond	40°	Remazol
Vinyl amide	Activated double bond	40°	Remazol

Reactive dyes are of outstanding importance for the dyeing of cotton, enabling bright intense coloration with high fastness. Approximately 95% of reactive dyes are azo dyes covering all the range of colours. Blues and greens are also provided by anthraquinone and phthalocyanine structures. As the name of these dyes suggests, they react with the fibre, whether cellulosic (cotton) or protein (wool) to form covalent bonds (Table 4). The two stages, first dyeing, then reaction, can take place separately or simultaneously. The characteristic structural feature is the presence of one or more reactive groups. Typically the dyes are depicted as:

D–B–RG

where D is the chromogen, B a bridging group and RG the reactive group. The most important reactive groups are the chlorinated triazines and vinylsulfones.

One of the three isomers of the simplest triazine is:

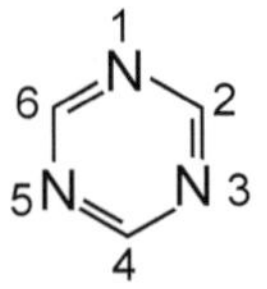

1, 3, 5-triazine

Figure 13.1

An example of a dye with a dichlorotriazine group is C.I. Reactive Blue 109:

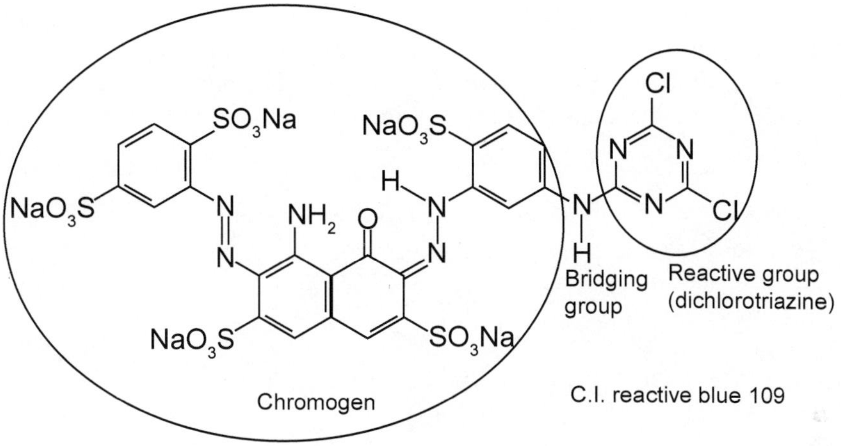

Figure 13.2

The reaction between the –OH groups of the cellulose in the fibre and the –C–Cl groups in the chlorotriazine is by a (nucleophilic) substitution reaction to form covalent bonds.

An ethenyl (vinyl) sulfone contains the $CH_2{=}CHSO_2$ group and the simplest is diethenylsulfone (divinylsulfone). The sulfone group can be seen in C.I. Reactive Blue 19:

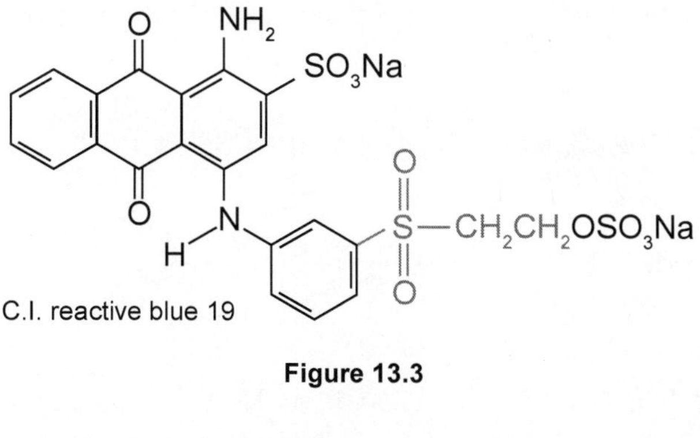

Figure 13.3

In this example, there is no bridging group. The dye reacts with cellulose by addition to the sulfur-oxygen double bond. Reactive dyes, in the aqueous solution, can undergo hydrolysis of the sulfone making it unreactive to the cellulose. This means that unreacted dye, if not washed off properly, will remain on the surface of the fabric giving an apparent colour that will wash out over time. To reduce this problem, dyes have been designed with two different reactive groups of differing reactivity. These dyes offer improved fastness because if one of the groups is hydrolyzed in solution, the other one will react with the hydroxyl groups in the fabric. The first of these included both a chlorotriazine and vinylsulfone groups and an example is C.I. Reactive Red 194:

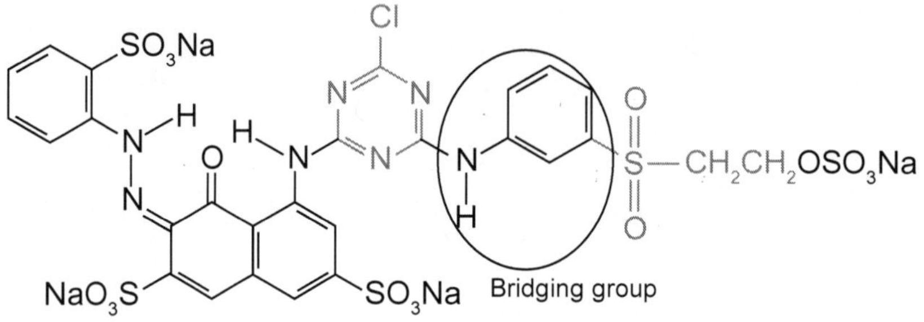

C.I. reactive red 194

Figure 13.4

Besides the two different reactive groups, there is a chromogen and a bridging group. All the reactive dyes have a relatively small molecular size and they also have two or more sulfonic acid groups in the chromogen, leading to a high solubility in water. A proportion of dye species (anionic) does not react with the fibre and is hydrolyzed and the product must be removed by washing.

13.6   Classification of reactive dyes (with example)

1. *On the basis of reactive group:* These are of two types
   • Halogen-added group (Cl2, F2, Br2, I2)
   Example: Pyrimidine
   • Vinyl-added group (–CH = CH2)
   (–D–SO$_2$–NH–CH = CH$_2$) [–CH = CH$_2$ … Reactive group]
   Example: Levafix
2. *On the basis of reactivity:* These are of three types
   • High reactivity, e.g. Procion – E [Medium alkali (NaHCO$_3$) used]
   • Moderate reactivity, e.g. Livafix – E [Medium alkali (Na$_2$CO$_3$) used]
   • Low reactivity, e.g. Premazine [Strong alkali (NaOH) used]

3. *On the basis of use:* These are of two types
    • Cold brand reactive dyes (High reactivity) – 40°C to 50°C
    • Hot brand reactive dyes (Low reactivity) – 90°C to 95°C
    • Use of hot brand is maximum in our country; cold brand is used for batik, tye dye, etc.

13.7   Advantages of the reactive dyes

•   Show improved fastness properties
•   Simplify dyeing procedure
•   Easy washability
•   Permanency of the color
•   Good chemical binding
•   Allows for a wide variety of chromophores to be used

13.8   Dyeing cycle and important factors/phases in reactive dyeing

•   pH of the substrate prior to dyeing
•   pH of the dye bath
•   Pretreatment of the substrate
•   Solubility of the dyestuff
•   Dyeing temperature
•   Quality of water and salt
•   Electrolyte concentration
•   Dyeing time
•   Washing off sequence
•   Type of alkali

13.9   Types of reactive dyes

Bifunctional dyes – Dyestuffs containing two groups are known as bifunctional dyestuffs. These reactive dyes are designed in such a manner to have the capacity to react with the fibre in more than a single way.

Vinylsulphone dye (VS) – Vinylsulphone dyes are moderately reactive. The dyeing temperature is generally 600°C and pH is 11.5 that gets applied by utilising a mixture of soda ash and caustic soda. These dyes show excellent fixation properties under proper alkaline condition. A typical example is the Remazol Black B (C.I. Reactive Black 5).

Monochlorotriazine dye (MCT) – Normally these dyes are less reactive than vinylsulphone dyes. Reaction can take place in more energetic reaction conditions. That is typically 800°C and pH value of 10.5 are essential for a proper fixation on cellulosic fibres. A typical monochlorotriazine dye is shown here.

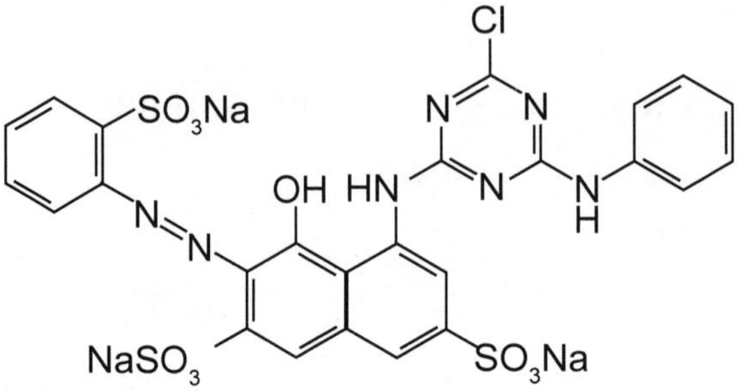

Figure 13.5

13.10   Uses of reactive dyes

By reactive dyes, the following fibres can be dyed successfully:
1. Cotton, rayon, flax and other cellulosic fibres
2. Polyamide and wool fibres
3. Silk and acetate fibres

13.11   Trade names of reactive dyes

Trade name	Manufacturer	Country
Procion	I.C.I	UK
Cibacron	Ciba	Switzerland
Remazol	Hoechst	Germany
Levafix	Bayer	Germany
Reactone	Geigy	Switzerland
Primazin	BASF	Germany
Drimarine	Sandoz	Switzerland

13.12   Properties of reactive dyes

1. Reactive dyes are anionic dyes, which are used for dyeing cellulose, protein and polyamide fibres.
2. Reactive dyes are found in power, liquid and print paste form.

3. During dyeing the reactive group of this dye forms covalent bond with fibre polymer and becomes an integral part of the fibre.

4. Reactive dyes are soluble in water.

5. They have very good light fastness with rating about 6. The dyes have very stable electron arrangement and can protect the degrading effect of ultra-violet ray.

6. Textile materials dyed with reactive dyes have very good wash fastness with rating. Reactive dye gives brighter shades and has moderate rubbing fastness.

7. Dyeing method of reactive dyes is easy. It requires less time and low temperature for dyeing.

8. Reactive dyes are comparatively cheap.

9. Reactive dyes have good perspiration fastness with rating 4–5.

10. Reactive dyes have good perspiration fastness.

13.13   General structure of reactive dyes

The general structure of reactive dye is D–B–G–X.

13.14   Chemical structure of reactive dyes

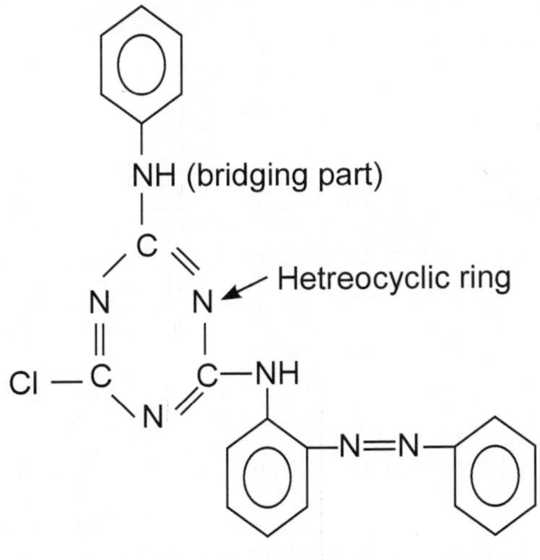

Figure 13.6 Chemical structure of reactive dyes

Here,

D = Dye part or chromogen (color producing part)

Dyes may be direct, acid, disperse, premetallised dye, etc.

B = Bridging part

Bridging part may be –NH– group or –NR– group.

G = Reactive group bearing part

X = Reactive group

13.15   Classification of reactive dyes

Reactive dyes may be classified in various ways as below:

(1) *On the basis of reactive group –*

    (a) Halogen (commonly chlorine) derivatives of nitrogen-containing heterocycle, such as

      • Triazine group

      • Pyridimine group

      • Quinoxaline dyes

      Example:

      Triazine derivatives: Procion, cibacron

      Pyridimine derivatives: Reactone

      Quinoxaline derivatives: Levafix

    (b) Activated vinyl compound:

      • Vinyl sulphone

      • Vinyl acrylamide

      • Vinyl sulphonamide

      Example:

      • Vinyl sulphone: Remazol

      • Vinyl acrylamide: Primazine

      • Vinyl sulphonamide: Levafix

(2) *On the basis of reactivity –*

    (a) Lower reactive dye/medium reactive dye: Here pH is maintained 11–12 by using Na_2CO_3 in dye bath.

    (b) Higher reactive dye: Here pH is maintained 10–11 by using $NaHCO_3$ in dye bath.

(3) *On the basis of dyeing temperature*

    (a) Cold brand – These types of dyes contain reactive group of high reactivity. So dyeing can be done in lower temperature i.e. 320–600°C. For example: Procion M, Livafix E.

(b) Medium brand – This type of dyes contains reactive groups of moderate reactivity. So dyeing is done in higher temperature than that of cold brand dyes i.e. in between 600°C and 710°C temperatures. For example, Remazol, Livafix are medium brand dyes.

(c) Hot brand – This type of dye contains reactive groups of least reactivity. So high temperature is required for dyeing i.e. 720–930°C temperature is required for dyeing. For example, Pricion H, Cibacron are hot brand dyes.

13.16   Manufacturing of reactive dyes

13.16.1   Reactive Red

Structure formula – The dicholoro-s-triazinyl dyes can be prepared by either of the following routes:

(1) Preparation of an azo dye containing a free amino group and then condensing the azo dye with cyanuric chloride.

(2) Coupling a diazotised primary aromatic amine with a cyanurated coupling component.

The choice of one of the above two routes depends mainly on the rate of the diazo-coupling reaction. If the coupling is very sluggish, route (1) is preferred.

Raw materials –

• Powdered Cyanuric Chloride
• H-Acid
• Sodium Carbonate (Soda Ash)
• Aniline
• Hydrochloric Acid
• Sodium Nitrite

Process details – Add 190 kg of finely powdered cyanuric chloride to a well-stirred mixture of crushed ice and ice cold water. Then gradually add a 1500 litres neutral solution of H acid and maintain the temperature below 5°C. Simultaneously add 20% (w/v) of sodium carbonate (soda ash) solution in small lots to maintain the pH between 6.2 and 6.8. The reaction mixture is stirred for 20–25 minutes and the pH is adjusted between 6.2 and 6.8 till the reaction is complete. This is indicated by the complete solubility of the reaction mixture. Absence of free diazotisable amino group also indicates the completion of the reaction.

Aniline is diazotized by the conventional method as follows:

About 260 litres (2.52 mole) of concentrated hydrochloride acid is added to a stirred mixture of 500 kg of ice and 300 litres of water and this is followed by the addition of 94 kg (1.01 mole) aniline. When aniline goes completely in the solution, add 70 kg (1.01 mole) sodium nitrite as 20% solution (w/v) to diazotized aniline. The temperature is maintained below 5°C and the completion of diazotization is tested with starch iodide paper. Then gradually add the diazo solution in a thin stream to the cyanurated H-acid. Simultaneously add 20% (w/v) sodium carbonate solution in lots to maintain the pH between 6.2 and 6.8. The completion of the coupling is indicated by the absence of the coupling component (here cyanurated H-acid) which is tested by spotting the reaction mixture and diazotized aniline side by side on a filter paper. The reaction is complete when there is no colour developed at the junction of the two run-outs of the spots. It generally took two and a half hours for completion of coupling. The temperature during coupling is kept below 10°C. A mixture of potassium dihydrogen phosphate and disodium hydrogen phosphate is then added as buffer. The quantity depends upon the final volume of the reaction mixture. About 3.4 g/l of potassium dihydrogen phosphate and 8.9 g/l of disodium hydrogen phosphate are found to be satisfactory. The reactive dye is then salted out with common salt, filtered and washed with chilled saturated brine containing the buffer and dried below 45°C under reduced pressure.

13.16.2   Reactive Orange

Structural formula – This dye is obtained by coupling diazotized meta-anilic acid with cyanurated J-acid.

Process details – The cyanuration of J-acid is carried out on similar lines as that of cyanuration of H-acid described in Reactive Red. Use of Acetone is not necessary. Diazotisation of metanilic acid is carried out by the conventional method and coupling is carried out at pH 6.2–6.8 and temperature 0–10°C. Rest of the operations are similar to those in the case of the above dyes. Salting out of Reactive Orange is easy as compared to the other dyes. The stirring during salting out stage is initially somewhat difficult but becomes smooth within a short time. The molar proportion again remains the same as in the case of Reactive Red.

13.16.3   Reactive Yellow

Structural formula – This dye is obtained by preparing the azo dye from diazotized C-acid and m-toluidine first and then cyanurating it in presence of acetone.

Process details – The diazotization of C-acid (1 mole) is carried out by the conventional method and the coupling with m-toluidine (0.99 mole) is carried out in acidic medium. The pH of the reaction mixture during coupling is found to have an appreciable effect on the colour yield of the product. Since the diazonium salt of C-acid has a very low solubility in highly acidic conditions, so either increase the volume for smooth and complete coupling or raise the pH to about 5.5 to 6.0. Raising the pH increases the solubility of the diazonium salt, coupling is more smooth and yield of the final dye is also increased. The cyanuration of the dye is carried out first by preparing a slurry of cyanuric chloride by using acetone as in the case of Reactive Red. Use of acetone reduces the time of cyanuration. The rest of the operations are similar to the ones used in the above dyes.

13.17   Identification of reactive dyes

In order to distinguish these dyes from other classes which color wool in the polyfibre test, an individual dyeing on mercerized cotton is necessary.

Procedure

Dissolve 150 mg of the dye sample in a 200 ml neutral dyebath at room temperature. Add 10 g of sodium sulphate, 5 g of mercerized cotton and 0.5 g of sodium carbonate. Bring to a boil within 10 min and continue boiling for 10 min. Rinse, then boil with synthetic detergent for 2 min. Only reactive and direct dyes remain on the fabric.

Dry the fabric and treat it with dimethylformamide. Direct dyes will be stripped; reactive dyes remain on the fabric.

Paper and thin-layer chromatography is very useful in the identification of reactive dyes.

13.18   Dyeing mechanism of reactive dye

The dyeing mechanism of material with reactive dye takes place in three stages:

1. Exhaustion of dye in presence of electrolyte or dye absorption.
2. Fixation under the influence of alkali.
3. Wash-off the unfixed dye from material surface.
4. Now they are mentioned below:

Dye absorption – When fibre is immersed in dye liquor, an electrolyte is added to assist the exhaustion of dye. Here NaCl is used as the electrolyte. This electrolyte neutralizes absorption. So when the textile material is introduces to dye liquor, the dye is exhausted on to the fibre.

Fixation – Fixation of dye means the reaction of reactive group of dye with terminal –OH or –NH_2 group of fibre and thus forming strong covalent bond with the fibre and thus forming strong covalent bond with the fibre. This is an important phase, which is controlled by maintaining proper pH by adding alkali. The alkali used for this creates proper pH in dye bath and do as the dye-fixing agent. The reaction takes place in this stage is shown below:

1. $D–SO_2–CH_2–CH_2–OSO_3Na + OH-Cell = D–SO_2–CH_2–CH_2–O–Cell + NaHSO_3$

2. $D–SO_2–CH_2–CH_2–OSO_3Na + OH-Wool = D–SO_2–CH_2–CH_2–O–Wool + NaHSO_3$

Wash-off – As the dyeing is completed, a good wash must be applied to the material to remove extra and unfixed dyes from material surface. This is necessary for level dyeing and good wash-fastness. It is done by a series of hot wash, cold wash and soap solution wash.

Application method – These are three application procedures available:

(1) Discontinuous method

   • Conventional method

   • Exhaust or constant temperature method

   • High temperature method

   • Hot critical method

(2) Continuous method

   • Pad-steam method

   • Pad dry method

   • Pad thermofix method

(3) Semi-continuous method

   • Pad roll method

   • Pad jig method

   • Pad batch method

13.19   Stripping of reactive dye

Stripping becomes necessary when uneven dyeing occurs. The reactive dye cannot be satisfactory stripped from fibre due to covalent bond between dye molecule and fibre. By stripping azo groups (–N = N–), brome the dye is removed. There are two methods of stripping:

13.19.1   Partial stripping

Partial stripping is obtained by treating the dyed fabric with dilute acetic acid or formic acid. Here temperature is raised to 70–100°C and treatment is continued until shade is product of hydrolysis. The recommended concentration is between 5 and 10 ml glacial acid or 2.5–10 parts of formic acid (85%) per 1000 parts of water. The amount of acid used is as below:

- Glacial acetic acid: 5–10 parts
- With water: 1000 parts
- Time: 20–30 min
- Temperature: 85°C to 90°C

Or

- Formic acid: 2.5–10 parts
- With water: 1000 parts
- Temperature: 70–100°C
- Time: Until desired shade is obtained

Full stripping: For complete stripping, the goods are first treated with sodium hydrosulphite ($Na_2S_2O_4$) at boil. Then washed off and bleached at ordinary atmospheric temperature in liquor containing 1 part per 100 of commercial sodium hypochlorite.

Recipe:

- $Na_2S_2O_4$ – 5 g/L
- Na_2CO_3 – 2 g/L
- Boiling – 20–30 min
- Time – 20–30 min
- Temperature – Boiling

The stripping recipe for reactive dyes is as follows:

- 15 ml/L Sodium hydroxide
- g/L Sodium hydrosulfite
- 2 g/L DA-BS800

Keep for 30–60 min at 60–80°C, then rinse thoroughly.

13.20   Different methods of reactive dye application

(1) Pad-batch method – Pad batch processes are of two types:
  (a) Pad (alkali)-batch (cold) process
  (b) Pad (alkali)-batch (warm or hot) process
(2) Pad dry method
(3) Pad steam method

13.21   Testing of reactive dyes

Recipe:

- H_2SO_4 – 1 cc/litre of water
- $Na_2S_2O_4$ – 2 cc/litre of water
- M: L – 1:20
- Temperature – Up to boiling
- Time – 30 min

When a sample of reactive dyed fabric is treated in a test tube containing H_2SO_4 of 1 cc per 1 litre water, it results bleaching of reactive color from the dyed fabrics. It is one of the identification tests of reactive color.

The reactive color remains fixed on the textile material though it is treated or boiled with pyridine or chloroform. On the other hand, textile material dyed with direct, azoic, etc., dyestuffs and treated with above chemicals then color will come out easily. It is one of the identification tests of reactive color. The second one is one of the most efficient test methods of reactive dye.

The Society of Dyers and Colourists and the American Association of Textile Chemists and Colorists, joint publishers of the Colour Index, are pleased to announce that Huntsman Textile Effects (TE) has registered its new products with the publication.

Colorants are identified worldwide by their Colour Index (CI) generic names. New names are assigned only for new structures and may be used to identify dyes and pigments from a variety of manufacturers.

Huntsman TE recently presented three of its newest products for registration. All three were recognized as unique structures and issued new generic names and numbers. By agreement with Huntsman TE, the structures are held confidentially by the Colour Index and will not be published.

Avitera Yellow SE is now also CI Reactive Yellow 217

Avitera Red SE is CI Reactive Red 286

Avitera Deep Blue SE is CI Reactive Blue 281

According to a Huntsman TE representative, "Huntsman TE has applied for the Colour Index numbers for their new Avitera SE reactive dye range, as they recognize the importance to have the Colour Index as a guide for available dyes in the world."

Only dyes registered with the Colour Index should use the CI designation. Dye users can check the Colour Index to ensure they are purchasing dyes that have been registered.

The Colour Index is the definitive guide to the technical details of dyes and pigments, and the companies that manufacture them. The fourth edition of the Colour Index is now available online and includes nomenclature, constitution, main applications, and suppliers. It is published jointly by the Society of Dyers and Colourists and the American Association of Dyers and Colorists.

The SDC is the leading independent, educational charity dedicated to advancing the science and technology of color worldwide. Its mission is to communicate the science of color in a changing world. The organization celebrated its 125th anniversary last year.

13.22   Current trends in reactive dyes and their application

- Warm dyeing application offers many advantages on the ground of utilisation of energy. So, there is a perceptible shift from hot dyeing application to warm dyeing.
- The chlorotriazine group together with vinyl sulphone has dominated on both technical & commercial grounds.
- Vat dyes are being replaced more & more with reactive dyes due to availability of new generation reactive dyes.
- The demand for Right-First Dyeing is increasing.
- The pressure on environmental issues and cost economics is increasing due to stiff competition in the global market.

13.23   Future trends in reactive dyes

Research is being carried out for:
- Increasing the robustness of individual dyes and dye combinations in trichromatic systems.
- Enhancing reproducibility of trichromatic combinations used in most commonly applied dyeing processes.
- Reducing salt consumption and/or unused dye in the effluent (Dyes with no salt, low alkali addition and 100% fixation).
- Improving fastness properties (e.g., light fastness, fastness to repeated laundering).
- Polyfunctional dye chemistry to improve reactivity, fixation levels and reproducibility.
- Reactive dyes exclusively for printings, which are different from low-molecular weight MCT dyes.

Reactive dyes are widely used for the dyeing of cellulose fibres and fabrics because of their economical dyeings and simple dyeing operating. The newer types of reactive dyes offer quite reasonable and accepted levels of wet and other fastness properties coupled with the cost advantages. These dyes have made vat dyes practically non-existent. Reactive dyes generally offer a wide range of bright colors, excellent performance, strong applicability, appealing hue and good performance on cellulosic fibers. Reactive dyes have received excellent market acceptability especially for the cellulosic fibre/fabrics. The global share of reactive dyes production is estimated to be 20~30% of the total dyestuff production. They are ranked first for application on cellulosics. But ordinary reactive dyes have poor utilization rate (poor exhaustion) mostly in the range of 60~65%. The rate of redyeing is high which increases the cost and total delivery time. The existing processes are not only time-consuming but are associated with high-energy consumption and huge amount of wastewater that is a lot more difficult to treat.

China uses approximately 10 million tons of reactive dyes annually. Experts estimate that if all change to use fluorine-containing reactive dyes or similar reactive dyes, the industry can reduce printing and dyeing wastewater by about 4,500 tons every year. The usage of inorganic salts can also be reduced by 3.25 million tons to 45 million tons. The use of fluorine containing reactive dyes enhances the fixation and a level of 80% exhaustion is achieved resulting in the benefits described above. But, the production of fluorine containing reactive dye series involves the use of very high technology, which is quite complex and challenging. Only a handful of global companies – DyStar in Germany, Huntsman and Clariant in Basel, Switzerland, have such manufacturing capabilities. The Color Root, therefore, offers an attractive alternative to developing countries with its award winning technology at much lower price. The technology was developed by the company in co-operation with the scientists at the Wuhan Textile University, a leading textile education and research centre.

13.24   Reactive dyes for wool

Lanasol CE dyes – The LANASOL range consists of sulfo group containing reactive dyes which have been especially developed for wool dyeing. They contain 1 or 2 bromoacrylamide reactive groups which form a covalent bond with the nucleophilic groups of the wool's amino acids during the dyeing process, resulting in outstanding wet fastness properties. Lanasol CE reactive wool dyes have been especially developed for cost effective deep shades and for replacement of afterchrome dyes. Lanasol and Lanasol CE dyes are applied from a weakly acid bath in the presence of Albegal B leveling agent. They are suitable for untreated and shrink resist treated, e.g. chlorinated and machine washable wool in all forms, especially loose stock, slubbing and yarn. They can also be used on silk.

13.24.1   Key features of Avitera SE dyes

- AVITERA® SE has three reactive groups in each dye molecule instead of the usual two, which allow dye to fix quickly to fibers with a very small gap between exhausted and fixed dyestuffs, ensuring a higher yield.
- In conventional dyes, 60–80 per cent of the dye applied to cotton during dyeing is fixed, but in AVITERA® SE, this is nearly 90 per cent.
- AVITERA® SE reactive provide an exceptionally low temperature, low resources consumption processing system for cellulosic fibers and their blends.
- AVITERA® SE has excellent solubility, high diffusion and outstanding washing-off properties, making them suitable for application at ultra-short liquor ratios. Their low sensitivity to variations in liquor ratio, exhaustion and fixation time and temperature ensure that reprocessing is reduced to a minimum.

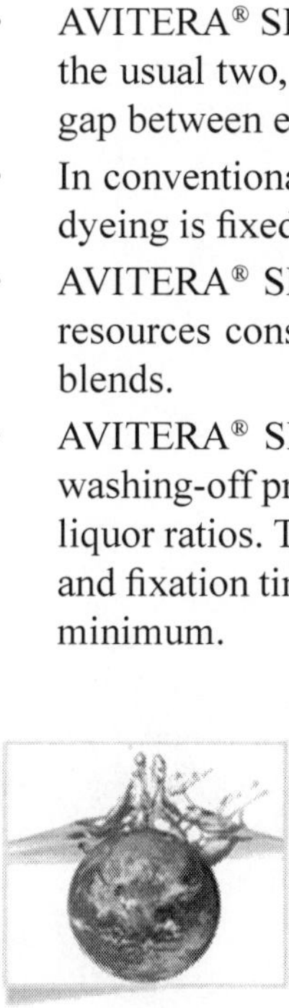

Dyeing process (ind. washing off.) for 1 kg cotton	Water (liter)	Energy* (kg steam)	Time (hour)
Conventional dyeing system	40.50	6.5	7
Best available technology (BaT)	15.20	1.7	4
AVITERA® SE	**15.20**	**1.7**	**4**

*Significant improvement in carbon footprint

Figure 13.7

Benefits
- As the temperatures never exceed 60°C and the number of rinsing baths is greatly reduced, AVITERA® SE can save more than 50 per cent of water, 70 per cent of energy and 50 per cent of time.
- The use of AVITERA® SE can reduce water consumption by 50%, meaning that only 15–20 liters of water are required to dye 1 kg of material.
- Predominantly, the cost saving is gained through a massive reduction in the amount of water and energy required for washing-off. With 5 per cent or less unfixed dye to be removed instead of the usual 15–30 per cent,

the washing-off process can be drastically curtailed and requires only a fraction of the water and energy needed by conventional dyes.

- Alongside AVITERA® SE – GENTLE POWER BLEACH™ could be used as a pretreatment to bring further improvement in results.

Figure 13.8

13.24.2   Fluorine-based reactive dyes

Later in 1980s with the introduction of Sumifix supra dyes by Sumitomo, textile processors have experienced a trend of incorporating different reactive groupings into the same molecule. These Hetero–Bi-functional Sumifix supra dyes contain both MCT and a Vinyl Sulphone reactive grouping. Due to presence of these two different reactive systems, these colours offered good fixation, over a wide range of dyeing temperature and being smaller in size, also offered better wash-off characteristics. Year 1990s witnessed the development of Fluorotriazine and low salt reactive systems, which resulted in excellent compatibility for achieving batch–to–batch reproducibility and use of low salt in dyeing. Fluorotriazine system offers many advantages over chlorotriazines. Traditional reactive dyes require the addition of a large amount of salt to achieve exhaustion. Salt not only facilitates the binding process of reactive dyes to cellulosic fibre, but also prevents the large-scale bonding of water molecules to the negatively charged dyes, which produces inert 'Dead dye'. This large amount of salt when discharged into bodies of water causes an increase in ecological salinity.

With the use of sophisticated molecular engineering techniques, it has been possible to design reactive dyes (e.g., low salt reactive dyes) with considerably higher performance than traditional reactive dyes. With the introduction of these revolutionary dyes, it is possible to reduce salt requirement by 50–60% based on the weight of fabric dyed. This was a very important development in the history of reactive dyes on the ecological point of view. Due to introduction of highly electrophilic 'F' group in the reactive system, more stable bond formed between cellulose and dye resulted into excellent overall fastness.

Some typical examples of reactive systems for cellulose fibres are reported below.

The late development of fibre-reactive dyes was partly caused by a lack of considerable reactivity of fibres made of cellulose or proteins. Despite the many possible reactive groups in dyes capable of covalent bond formation with nucleophilic groups in wool, only a limited number of types of reactive dye have been commercially successful. Some typical examples of reactive systems for wool or polyamide fibres are reported below.

The most important reactive groups in wool are all nucleophilic and found mainly in the side-chains of amino acid residues. They are in order of decreasing reactivity: thiol, amino and hydroxyl groups. The actual dyes are probably dibromopropionamides, which eliminate HBr on dissolving in hot water. Mythyltaurineethylsulphones and 2-sulphatoethylsuphones form the vinyl sulphone reactive group relatively slowly at pH 5–6. This allows some levelling during dyeing before the vinylsulphone dye reacts with the wool and becomes immobilised. Bromoacrylamido groups are stable in boiling water at pH 7 and react by both nucleophilic addition and nucleophilic substitution reaction.

Index